SITE INVESTIGATIONS FOR REPOSITORIES
FOR SOLID RADIOACTIVE WASTES
IN SHALLOW GROUND

The following States are Members of the International Atomic Energy Agency:

AFGHANISTAN	HOLY SEE	PHILIPPINES
ALBANIA	HUNGARY	POLAND
ALGERIA	ICELAND	PORTUGAL
ARGENTINA	INDIA	QATAR
AUSTRALIA	INDONESIA	ROMANIA
AUSTRIA	IRAN	SAUDI ARABIA
BANGLADESH	IRAQ	SENEGAL
BELGIUM	IRELAND	SIERRA LEONE
BOLIVIA	ISRAEL	SINGAPORE
BRAZIL	ITALY	SOUTH AFRICA
BULGARIA	IVORY COAST	SPAIN
BURMA	JAMAICA	SRI LANKA
BYELORUSSIAN SOVIET	JAPAN	SUDAN
SOCIALIST REPUBLIC	JORDAN	SWEDEN
CANADA	KENYA	SWITZERLAND
CHILE	KOREA, REPUBLIC OF	SYRIAN ARAB REPUBLIC
COLOMBIA	KUWAIT	THAILAND
COSTA RICA	LEBANON	TUNISIA
CUBA	LIBERIA	TURKEY
CYPRUS	LIBYAN ARAB JAMAHIRIYA	UGANDA
CZECHOSLOVAKIA	LIECHTENSTEIN	UKRAINIAN SOVIET SOCIALIST
DEMOCRATIC KAMPUCHEA	LUXEMBOURG	REPUBLIC
DEMOCRATIC PEOPLE'S	MADAGASCAR	UNION OF SOVIET SOCIALIST
REPUBLIC OF KOREA	MALAYSIA	REPUBLICS
DENMARK	MALI	UNITED ARAB EMIRATES
DOMINICAN REPUBLIC	MAURITIUS	UNITED KINGDOM OF GREAT
ECUADOR	MEXICO	BRITAIN AND NORTHERN
EGYPT	MONACO	IRELAND
EL SALVADOR	MONGOLIA	UNITED REPUBLIC OF
ETHIOPIA	MOROCCO	CAMEROON
FINLAND	NETHERLANDS	UNITED REPUBLIC OF
FRANCE	NEW ZEALAND	TANZANIA
GABON	NICARAGUA	UNITED STATES OF AMERICA
GERMAN DEMOCRATIC REPUBLIC	NIGER	URUGUAY
GERMANY, FEDERAL REPUBLIC OF	NIGERIA	VENEZUELA
GHANA	NORWAY	VIET NAM
GREECE	PAKISTAN	YUGOSLAVIA
GUATEMALA	PANAMA	ZAIRE
HAITI	PARAGUAY	ZAMBIA
	PERU	

The Agency's Statute was approved on 23 October 1956 by the Conference on the Statute of the IAEA held at United Nations Headquarters, New York; it entered into force on 29 July 1957. The Headquarters of the Agency are situated in Vienna. Its principal objective is "to accelerate and enlarge the contribution of atomic energy to peace, health and prosperity throughout the world"

Printed by the IAEA in Austria
April 1982

TECHNICAL REPORTS SERIES No. 216

SITE INVESTIGATIONS FOR REPOSITORIES FOR SOLID RADIOACTIVE WASTES IN SHALLOW GROUND

INTERNATIONAL ATOMIC ENERGY AGENCY
VIENNA, 1982

*A publication within the IAEA programme on
the Underground Disposal of Radioactive Wastes*

SITE INVESTIGATIONS FOR REPOSITORIES
FOR SOLID RADIOACTIVE WASTES IN SHALLOW GROUND
IAEA, VIENNA, 1982
STI/DOC/10/216
ISBN 92–0–125382–6

FOREWORD

This report is essentially a companion document to Safety Series No.53, "Shallow Ground Disposal of Radioactive Wastes: A Guidebook", published by the IAEA in 1981. It is addressed to administrative and technical authorities responsible for or involved in planning, approving, executing and reviewing national waste-management programmes, and is expected to contribute to an understanding among the several disciplines involved in the necessary investigations.

A Consultants' Meeting was convened in Vienna from 28 November to 2 December 1977 to make recommendations on the preparation of technical and safety reports concerned with the earth science and other studies required for the selection and confirmation of underground repository sites. Based on these recommendations, a working paper for this report was drafted at a Consultants' Meeting held in Prague from 23 to 27 July 1979. The working paper was reviewed and revised at an Advisory Group Meeting held in Vienna from 10 to 14 December 1979 and later by a Consultant in Vienna from 17 to 19 December 1979. The report was finally examined by the Technical Review Committee on the Underground Disposal of Radioactive Waste at a meeting in Vienna from 10 to 14 November 1980.

The IAEA has been active in the field of radioactive waste management for many years. In 1977, a draft proposal was prepared for a future integrated IAEA programme on the disposal of radioactive waste into geological formations. An Advisory Group Meeting held from 30 January to 3 February 1978 confirmed the proposal and made recommendations regarding a programme to develop technical reports and guidelines in the field of underground disposal of radioactive waste. This programme is intended to cover the needs and interests of both developed and developing nations, and will include the following general subject areas:

Generic and regulatory activities and safety assessments
Investigation and selection of repository sites
Waste acceptance criteria
Design and construction of repositories
Operation, shut-down and surveillance of repositories.

The present publication is part of the IAEA's programme to develop guidelines and technical reports on these subjects.

Approaches and methods for site investigations have similarities according to the various disposal options. The specifics and the relative emphasis in investigations vary, however, and this requires the preparation of reports on particular subjects, such as this one, pertaining to shallow-ground disposal.

A number of related publications have been or are being prepared within the IAEA's programme on the safe underground disposal of radioactive wastes, dealing with possible options for the disposal of high-, intermediate- and low-level radioactive wastes in deep continental geological formations, in rock cavities at various depths and in shallow ground. Publications in this series are as follows:

Site Selection Factors for Repositories of Solid, High-Level and Alpha-Bearing Wastes in Geological Formations, IAEA Technical Reports Series No.177 (1977)

Development of Regulatory Procedures for the Disposal of Solid Radioactive Waste in Deep, Continental Formations, IAEA Safety Series No.51 (1980)

Shallow Ground Disposal of Radioactive Wastes: A Guidebook, IAEA Safety Series No.53 (1981)

Underground Disposal of Radioactive Wastes: Basic Guidance, IAEA Safety Series No.54 (1981)

Safety Assessment for the Underground Disposal of Radioactive Wastes, IAEA Safety Series No.56 (1981)

Site Investigations for Repositories for Solid Radioactive Wastes in Deep Continental Geological Formations, IAEA Technical Reports Series No.215 (1982)

Other appropriate publications prepared under the Radiological Safety Standards Programme and the Nuclear Safety Standards (NUSS) Programme might be consulted under the various related topics.

The IAEA gratefully acknowledges that the preparation of this publication was partially funded by the United Nations Environment Programme (UNEP) under its Project No. 0102–74–002 with the IAEA.

The Agency wishes to express its thanks to all those who took part in the preparation of the report. The responsible officer at the IAEA for this report was K. Schneider, from the Waste Management Section of the Division of Nuclear Fuel Cycle. The work was completed with the assistance of K.T. Thomas of the same division.

CONTENTS

1. INTRODUCTION

Nuclear energy is playing a continuing and increasing role in the power economy of many countries. These countries must establish appropriate systems for the safe management of the radioactive wastes resulting from the use of nuclear energy.

The radioactive materials no longer useful to man become wastes that must be kept away from man for the duration of their potentially harmful levels of radioactivity. The radioactivity (and thus the radiotoxicity) in these wastes eventually decays with time to lower levels at rates depending on the half-lives of the specific radionuclides and their daughter products[1]. For most radionuclides of importance in management of wastes by shallow-ground disposal, the half-life varies from about one month to several decades. The transuranic content should not be significant[2]. Thus, the potential radioactivity hazard can exist for several hundred years, and it is incumbent on all nations producing radioactive wastes to prevent releases of amounts of the radioactivity into man's environment that could become harmful to man.

The potential impacts of these wastes on man and his environment should be kept acceptably low. The confinement of radioactive wastes should remain effective until the radionuclides have decayed to an acceptable level before they reach man's environment. Releases of radionuclides from the buried wastes are generally considered to be acceptable provided their radiological impacts on man conform to the basic principles of radiation protection as defined by the International Commission on Radiological Protection and relevant national authorities. With a sufficient number of natural and man-made barriers, the release of radioactive material can be delayed, its transport retarded or its concentration diluted enough to ensure that the impact on man will remain within acceptable levels.

Underground isolation of radioactive waste is the disposal option that at present appears to be the most viable and is receiving the most attention. For appropriately immobilized low- and intermediate-level and short-lived solid wastes, disposal into a repository constructed in shallow ground is one of the options being used to provide protection for man and his environment. A repository site typically includes a buffer zone round the waste disposal area as well as the repository itself where the wastes are emplaced.

Studies on disposal systems and the search for potential repository sites are important components of national nuclear power programmes. Investigations for

---

[1] The specific activity and/or radiotoxicity of some radioactive daughter nuclides are higher than that of the respective precursor; for this, the radioactivity level and/or the radiotoxicity level will increase for some time before it decreases.

[2] 'Not significant' or 'insignificant' here indicates that these amounts can be ignored for purposes of waste disposal (see Ref [1]).

repository sites are directed to producing confidence that the behaviour of the geological environment will provide appropriate confinement of the waste over the long period of concern. Investigations are usually directed at first to selection of favourable geological environments containing appropriate host geological materials. Subsequent site investigations are required to confirm that disposal of the radioactive wastes into or onto the geological formations at a specific site is feasible and can meet safety criteria and licensing requirements.

2. SCOPE

This report provides an overview and technical guidelines for investigations on a national level for the selection and confirmation of a repository site that will provide adequately safe performance for disposal of solid radioactive wastes that are low- or intermediate-level and short-lived. It also provides basic information on technical activities to be undertaken and on techniques that are available for such investigations in the various steps in selecting suitable sites. The report supplements the information given in *Shallow Ground Disposal of Radioactive Wastes: A Guidebook,* IAEA Safety Series No. 53 (1981).

While the general approach to and methods of site investigation are similar in the various disposal options, their emphasis and details on specifics vary according to the type and end use of each report. Certain information on matters of common interest has been taken from the following IAEA publications:

Site Selection Factors for Repositories of Solid High-Level and Alpha-Bearing Wastes in Geological Formations, IAEA Technical Report No.177 (1977).

Underground Disposal of Radioactive Wastes: Basic Guidance, IAEA Safety Series No. 54 (1981.

Site Investigations for Repositories for Solid Radioactive Wastes in Deep Continental Formations, IAEA Technical Report No. 215 (1982).

This was necessary in order to ensure that the present report can stand on its own and be understood in its entirety. IAEA Technical Report No. 215 is a comprehensive basic document on site investigation. The present report contains significant differences on important topics which are relevant to shallow-ground disposal.

In this document, shallow-ground disposal refers to the emplacement and covering of radioactive waste, with or without added engineered barriers, above or below the ground surface. Disposal of wastes in underground cavities may also be considerd a form of shallow-ground disposal if the cavity is sufficiently close to the surface, but this is outside the scope of the present report. The radioactive wastes considered here are those produced by man in various nuclear activities,

2

containing relatively small amounts of radionuclides and radionuclides that will decay to acceptable levels during the time that institutional controls for the repository are expected to last. The report does not deal with radioactive tailings from mining and milling of naturally occurring uranium and thorium ores.

The technical information in this report has been made general enough to cover a broad range of situations. Modification may, however, be necessary to reflect local conditions including variations in amounts and packaging of wastes. On the other hand, while it is believed that most major considerations are included, the information is not expected to be all-inclusive. The technical approach to site investigations and site selection as described here is generally applied to a highly complicated case in order to provide background information. In many specific cases, all the details given here may not be applicable to the extent implied. The report is based on available technology for site investigations.

This report focuses mainly on different aspects of earth sciences and the various investigative techniques relative to earth sciences that may be necessary for site investigations. Some major related studies in other fields are discussed briefly. It is assumed that no previous investigations have been undertaken, and the report proceeds through area site selection to the stage when the site is confirmed as suitable for a waste repository.

3. WASTE AND DISPOSAL CONCEPTS

Radioactive wastes are currently generated from two major sources: the different steps of the nuclear fuel cycle and the production and use of various radioisotopes. At present, the wastes produced from nuclear power production and its associated activities are generally greater in volume than those from isotope production and applications. While high-level and long-lived wastes are obviously out of the question for shallow-ground disposal, the suitability of a host of other types of wastes from various stages in the nuclear fuel cycle and from other radioisotopic sources needs to be carefully examined before being accepted for shallow-ground disposal.

An underground waste repository should perform two important and related functions: one is to isolate the waste in order to limit the entry of waste radio-nuclides into man's environment to acceptable levels; the other is to protect the waste from exposure to the dispersive effects of near-surface processes. Total containment, i.e. the assumption that there will never be a return of radionuclides to man's environment, may not be a realistic concept. Nevertheless, by emplace-ment in a repository designed with a sufficient number of natural and/or man-made

TABLE I. TENTATIVE RELATIONSHIP OF PREFERRED DISPOSAL OPTIONS AND RADIOACTIVE WASTE CATEGORIES

Disposal options		WASTE CATEGORY				
		High-level, long-lived	Intermediate-level, long-lived	Low-level, long-lived	Intermediate-level, short-lived	Low-level, short-lived
Emplacement in deep geological formations[c]	Dry[a]	Solid, immobilized, packaged, spaced for heat dissipation	Solid, immobilized, packaged	Solid, immobilized, packaged	Applicable, but may be more stringent than necessary[e]	Applicable, but may be more stringent than necessary[e]
	Wet[b]	As above; possibly with more engineered barriers	As above; possibly with more engineered barriers	As above; possibly with more engineered barriers		
Emplacement in mines or cavities[d]	Dry[a]	Not recommended	Possible, depending on circumstances	Possible, depending on circumstances	Solid, may be packaged[e]	Solid, may be packaged[e]
	Wet[b]	Not recommended	Not recommended	Not recommended	Solid, immobilized, packaged[e]	Solid, immobilized, packaged[e]
Emplacement at shallow depths	Dry[a]	Not recommended	Not recommended	Not recommended	Solid, immobilized, packaged	Solid, may be immobilized or packaged
	Wet[b]	Not recommended	Not recommended	Not recommended	Possible; immobilized, packaged, with more engineered barriers	Possible; solid, immobilized or packaged, with more engineered barriers
Injection of self-solidifying fluids into induced fractures in low-permeability strata		Not recommended	May be possible with adequate demonstrated technology and certain radionuclides	May be possible with adequate demonstrated technology and certain radionuclides	Applicable with appropriate technology	Applicable with appropriate technology
Liquid injection into deep, permeable formations		Not recommended	May be possible with adequate demonstrated technology and certain radionuclides	May be possible with adequate demonstrated technology and certain radionuclides	Applicable with appropriate technology	Applicable with appropriate technology

a Geological environments naturally isolated from moving groundwater.

b Geological environments with some movement of groundwater.

c Repository excavated especially for the disposal of radioactive wastes.

d The mine or cavity may have resulted from natural processes or from extraction of minerals, or may be excavated especially for waste disposal.

e May be preferred for countries with undesirable geological conditions at shallow depths.

barriers, the entry of the radioactive materials into the human environment can be limited. These barriers can be:

(a) **Natural barriers**

The geological formation in which the repository is sited, and the surrounding environment.
The retention along the possible pathways from the repository through the geological medium to the human environment.

(b) **Man-made barriers**

The physicochemical form of the waste (low leachability and low dispersibility).
The resistance of the container(s) to corrosion.
Additional engineered barriers, including geochemical and low-permeability barriers, in the repository.

3.1. WASTES SUITABLE FOR DISPOSAL

Radioactive wastes have many different characteristics. They contain varying amounts of radionuclides (ranging from trace quantities to about 10 wt% with different half-lives and toxicities[3], and varying amounts of bulk materials.

3.1.1. Radionuclide content

Radioactive wastes are divided into five generic categories for purposes of waste disposal; Table I summarizes their applicability to underground disposal [1]. For disposal purposes the wastes are categorized as having high, intermediate, or low levels of radioactivity, and short (usually less than about 5 to 30 years) or long half-lives. The category of waste content considered in this report, given as the last two columns in Table I, gives shallow ground as suitable for disposal of short-lived and intermediate- or low-level radioactive wastes after they have been put into solid form. For some wastes, immobilized waste forms and/or packaging is recommended.

To be suitable for shallow-ground disposal, the amount and/or concentration of radionuclides in the wastes should be such that radioactive decay will reduce the radionuclide content to levels acceptable for safety reasons during the period that institutional surveillance is expected to last. Because of differences in radioactivity decay rates, radiotoxicity, and radionuclide transport rates in the disposal system, the acceptable levels for shallow-ground disposal will differ for

---

[3] Some wastes, such as spent nuclear fuels, have significantly greater than 10 wt% radionuclides.

each of the radionuclides present. Thus, other conditions being equal, the acceptable short-lived radionuclide content will be higher than the longer-lived radionuclide content. To meet nuclear safety requirements, national authorities need, in practice, to establish appropriate limits for different groupings of radionuclides present in the different types of wastes for shallow-ground disposal.

3.1.2. Physical and chemical characteristics

Radioactive wastes occur in a variety of different physical and chemical forms. They contain varying amounts of materials of different chemical composition, which may not be degradable. In some cases, these characteristics of the bulk material make it necessary to convert the waste to another more stable form that is more suitable for disposal before being placed in its final repository. It is therefore recommended that radioactive wastes for disposal have good chemical, mechanical, biological, thermal and radiation stability, and low content of non-degradable toxic chemicals. Low mobility of the contained radionuclides is important for the period of concern.

Waste form stability is desired in order to minimize the possible effects of events that may occur during the required confinement time and that may tend to release the radionuclides into man's environment. Chemically undesirable characteristics include the presence of strong oxidants or corrosive agents, highly unstable chemicals, chemical complexing agents, etc.

Solid waste forms enhance most of the desired characteristics, and it is therefore considered that all wastes for shallow-ground disposal should be in solid form. In addition, it is preferred that this solid form should have low porosity, low surface area and low leachability in order to further reduce the potential mobility of the radionuclides contained therein. This suggests the high desirability of waste forms that are rigid monolithic solids.

3.1.3. Conditioning

The form in which some wastes arise may have desirable properties for shallow-ground disposal. On the other hand, it is frequently necessary to condition the waste to improve its suitability for disposal. Conditioning may be undertaken to reduce the volume and porosity of the waste, to convert from gaseous or liquid forms to solid forms in order to improve the stability and immobility of the waste and/or to provide appropriate packaging. Some of the treatment or conditioning methods available are: incineration of combustible materials; mechanical compaction of materials with high void content; precipitation or ion exchange from liquids or gases; evaporation to reduce volumes; and incorporation of wastes into monolithic blocks of concrete, bitumen or plastics; and packaging [2].

Stable and long-lasting packaging of the solid waste form, which is part of waste conditioning, will further enhance the desired properties of the waste disposal system. The packaging material and its design should have the same desirable characteristics as those discussed above for the waste form itself. Some waste forms may be acceptable without separate packaging, but proper packaging will usually improve the safety of the disposal system. In many cases, packaging will be required to provide safety during transport of the waste to the disposal site and during its emplacement and disposal.

3.2. SHALLOW-GROUND DISPOSAL PRACTICE

Shallow-ground disposal of radioactive wastes has been practised for several decades with wide variations in the procedures employed, the quantities and the types of low- and intermediate-level wastes disposed of. The greater part of radioactive wastes have been disposed of by this method.

Designs of shallow-ground disposal facilities have undergone extensive evolution as a result of operational experience and requirements to improve radiological safety and economy. This evolution has resulted in increasing the number of natural and engineered barriers, in limiting or delaying the migration of radionuclides from the repository, and in increased knowledge of the safety factors for waste disposal. Experience has shown that conditioned wastes not acceptable for disposal into a site with moderately favourable characteristics may be acceptable for highly favourable sites.

Overall, the important features for achieving radiological safety for shallow-ground disposal include (a) favourable site characteristics and (b) use of engineered barriers and appropriate wastes with suitable conditioning. Most current designs of repository systems include a number of these favourable characteristics.

Shallow-ground disposal technique consists in emplacement of wastes in disposal units constructed on, or in the vicinity of, the earth's surface. Emplacement of wastes into natural or artificial cavities at depths of several tens to perhaps hundreds of metres is not considered to be shallow-ground disposal in this report. This latter category requires significantly different emphasis in siting, designing and operating, and it is therefore outside the scope of this guide.

The oldest and simplest method of shallow-ground disposal is that of placing untreated solid wastes directly on the earth's surface and later covering the wastes with a layer of soil. The engineered barriers are minimal, primary protection being provided by the packaging and by the sorption and mechanical properties of the overlying and surrounding soil. Erosion, intrusion by animals, percolation of precipitation water, movement of groundwater, and other processes can adversely affect the safety of this method. Relatively large areas are required by this method, and it changes the appearance of the landscape. It is seldom used for new repositories for radioactive wastes because of its safety implications.

A related method is the emplacement of wastes into trenches dug in specific areas and subsequently covering the wastes with a layer of soil. The main disadvantages of the preceding method remain, relatively unchanged; the advantages are the requirement of a somewhat smaller area owing to the greater thickness of emplaced wastes, and the correspondingly smaller changes in the landscape.

Recent practice has been to locate trenches typically above the groundwater level and sometimes on a layer of clay or other material with low permeability and good sorption characteristics. These trenches are sometimes lined with bitumen or other material to improve confinement. The space between the waste packages can be filled with soil or other material with good sorptive properties. After being filled, the trench is covered by soil and rock rubble or by concrete panels. (In the latter case, the crevices between the panels can be filled with concrete or another sealant.) The whole surface is then usually covered by a layer of soil and material with low permeability. In this basic concept, three engineered barriers are applied (in addition to those associated with the waste condition): (1) the most important, materials with low permeability above the trench to minimize surface water entering the trench; (2) a water diversion and drainage system to direct water from the surface away from the trench; and (3) a trench bottom with low permeability. The system can be further protected from erosion by planting vegetation or covering with rock rubble and by careful contouring of the final surface. The space round the trenches, in turn, can be filled with gravel or sand or with another material to drain precipitation water away from the wastes.

In some countries, wastes are combined with protective materials in monolithic blocks near the earth's surface. For example, a concrete pit (which may be additionally lined internally and/or externally to reduce precipitation and the entry of groundwater) is filled with solid wastes, after which the free space is filled with a concrete mixture that may or may not contain radioactive concentrates. A layer of concrete is also poured over the top of the wastes.

These better practices are designed to improve the water management, e.g. to circumvent the possible 'bathtub effect'[4]. This is done largely by taking measures to ensure negligible penetration of water into the disposal area by using materials with low permeability above and beside the wastes. Material of low permeability is generally used also in the bottom liner of a repository, but acceptable safety may be possible without such a bottom liner, since control of water infiltration is of primary importance.

---

[4] The 'bathtub effect' is the accumulation of water in a shallow-ground facility. Typically, the facility does not adequately prevent water from entering the repository and has a bottom of low-permeability material that does not allow egress of water.

4. SHALLOW-GROUND DISPOSAL OBJECTIVES
AND SITING CONSIDERATIONS

Shallow-ground disposal of radioactive waste has been practised for a number of years. This experience has provided evolution and refinement of objectives, principles and site-selection factors, which are discussed in this section.

4.1. OBJECTIVES

The overall objective of shallow-ground disposal is to dispose of radioactive wastes in such a manner that any resultant radiation dose to man is within acceptable levels. To accomplish this, the following criteria should be satisfied:

(a) Man should be protected from unacceptable radiological effects of disposed wastes. Safety evaluations for potential disposal sites should take into consideration and evaluate the potential effects on man. Recommendations developed by the International Commission on Radiation Protection [3] should be implemented in accordance with national requirements.

(b) A disposal site should provide a high assurance for reliable prediction of a satisfactory safety performance. This implies in general that the geological/hydrological system should be relatively well understood and amenable to quantitative analysis.

(c) The disposal site should be compatible with the characteristics of the wastes it contains and with the repository design to ensure that the site, the waste form, and the engineered barriers provide adequate safety. The characteristics of these components should preferably be complementary.

(d) The known potential resources or potential land uses should not conflict unacceptably with use as a repository. They should also be assessed with respect to their long-term geologic stability and the possible consequences of climatological changes.

(e) Because shallow-ground disposal of radioactive wastes is a system wherein the hazardous materials are relatively accessible to man, institutional controls should be used to prevent undesired access by man or animals during the time that the potential hazard exists. Such necessary controls and surveillance measures could be more effective and would generally provide for improved safety over the long term if the number of disposal sites were kept to a practicable minimum.

4.2. SITING

When properly implemented, shallow-ground disposal is expected to provide adequate confinement of radionuclides for reasonable periods of time. This

TABLE II. MAJOR SITE SELECTION FACTORS

---

1. Topography

2. Tectonics and seismicity

3. Subsurface conditions:
   Depth of disposal zone
   Formation configuration; thickness and extent, consistency, uniformity,
      homogeneity and purity of strata
   Nature and extent of overlying, underlying and flanking beds

4. Geologic structure of the site area:
   Dip or inclination
   Faults and joints in nearby strata
   Diapirism in nearby strata

5. Physical and chemical properties of host material:
   Permeability, porosity, solubility and dispersivity
   Inclusions of gases and liquids
   Mechanical and plastic behaviour
   Sorption capacity
   Mineral content of water

6. Hydrology and hydrogeology:
   Surface waters: occurrence, form volumes
   Groundwaters: occurrence, volumes, chemistry

---

confinement may, however, be limited by the intrusion of man or animals and by certain natural processes. Shallow-ground disposal is therefore considered suitable for radioactive wastes that decay to acceptable levels within the period for which institutional control of the disposal site can be reasonably expected to last.

The objective of site selection is to ensure that the site has natural properties which provide adequate confinement of radionuclides from the human environment in concert with the engineered barriers in the repository. If repository barrier(s) fail and result in releases of radionuclides from the disposal system, the site characteristics should provide sufficient barriers to keep the radiological impact on man within acceptable levels.

An essential property of a shallow-ground repository is that the selected site be located in an area of favourable geological and hydrogeological characteristics such that the wastes, once emplaced, will be adequately isolated from the human

7. Future natural events:
   Hydrological changes
   Uplifts and subsidence
   Seismic events
   Intrusions and faulting
   Climatological changes
   Topographical changes

8. General geological and engineering conditions:
   Site area and buffer zone
   Pre-existing excavations
   Exploration excavations
   Spoil disposal
   Waste transport
   Engineering and construction of repository
   Operational safety and stability of repository

9. Societal considerations:
   Resource potential
   Land value and use
   Population distribution
   Jurisdiction and rights of the land
   Accessibility and services
   Other environmental impacts
   Public attitudes

environment for the relevant period. The characteristics of the disposal site, its location, and the design of the disposal facilities will determine the type, quantity and conditioning of wastes that can be emplaced. Releases of radionuclides to the environment from a disposal site may be acceptable provided the radiological impact conforms to basic radiological protection criteria.

The characteristics of the geological environment should be evaluated at the national, regional and site-specific levels to permit the selection of a site for a shallow-ground repository. Site selection factors must be taken into account in order to enable the identification of specific study areas which can then be subjected to further investigation. Site selection factors enter the process in at least two stages, namely the identification of areas worthy of more concerted study, and the evaluation of sites that emerge as having potential for the location of a shallow-ground repository.

11

Mechanisms for transport of radionuclides away from a disposal site are related to hydrological, geological, ecological and biological processes, and human and animal action. The hydrological characteristics of a disposal site are usually the main factors controlling the movement of radionuclides, since water is generally the most likely natural medium for off-site transport of radionuclides. Where hydrogeological characteristics are less favourable than desired, it is often possible to provide engineered means to enhance confinement. The characteristics of the disposal site and its supplementary engineered features provide the primary safety barriers for protection of man over the period of concern. It is also advantageous for the area surrounding the immediate repository ('buffer zone') to have characteristics that further enhance the safety of the site.

Owing to the predominance of parameters which are highly site-specific and interactive with each other, it is beyond the scope of this report to attempt to offer specific guidelines that will finally decide the selection of individual sites. However, general factors can be identified that can govern the suitability of potential sites as repositories. By identification of these factors and the manner in which they may affect the safe use of a site, general guidance can be developed for suitable repository sites. Major site selection factors for repositories for solid high-level and alpha-bearing wastes in geological formations are described in IAEA Technical Report No. 177 [4]. A modified list of these major factors is given in Table II. The relative position of the factors does not imply an order of priority since their relevance to the site selection process can vary in specific cases; for example, engineering and societal considerations can in many cases be the most important.

It is unlikely that any site will be found that incorporates all the advantageous factors. Neither is this necessary. The acceptance or rejection of a site will inevitably involve a balance between the advantageous and the less desirable factors. It is essential that the long-term safety requirements are satisfied and, to ensure this, all the confinement factors, natural and man-made, should be analysed.

5. OVERVIEW OF INVESTIGATIONS FOR SITE SELECTION

The term *site investigations* encompasses those fields of activities that are conducted in the course of evaluating a site as to its suitability for the location of a waste repository. The site investigations become more detailed as the site selection process continues, and culminate in site confirmation at one or more suitable sites. Site confirmation takes place simultaneously with the construction of the repository.

Selection of appropriate sites for shallow-ground disposal of solid radioactive wastes involves integration of site investigative work from a number of disciplines.

These involve many branches of earth sciences, engineering, safety analysis, health physics and social sciences. The investigations include theoretical, laboratory and field activities carried out in a generally stepwise fashion but with significant interaction between steps.

An idealized general sequence of activities for a relatively complex site selection procedure is shown in simplified schematic form in Fig.1 together with some of the major activities for each investigative stage[5]. The investigations start with an input of an identified need for a repository and conclude with the selection of a site or sites that are confirmed as meeting the safety requirements by detailed studies. The four principal steps in the process of site confirmation are shown in the boxes in Fig.1. The steps are intended to be generic but it should be emphasized that not all nations may wish to move sequentially from Step 1 through Step 4. Furthermore, the site investigations can often be much simpler than implied in Fig.1, especially when extensive engineered barriers are planned. The depth of activities in any one step will differ in different cases. Some studies undertaken as part of each step will overlap in time for most cases, and significant iteration and rework will be involved. The level of activities in Steps 3 and 4 is likely to be far greater than in Steps 1 and 2.

Engineering design studies of the repository are undertaken simultaneously with the site selection investigations, and provide important interactions with site selection factors. These studies include activities such as identification of the quantities and characteristics of the wastes, the waste forms and their packaging. At each step of the site investigations, societal, ecological and national legislative issues are considered and dealt with according to national policies. At relevant stages of the investigations, appropriate regulatory bodies should be kept informed and will be involved in decisions.

Before the earth-science programmes shown in Fig.1 are developed, it will be necessary for the national authorities to develop a master plan for the needs and timing for a repository, primarily as dictated by national waste arisings.

The investigations begin with desk studies and proceed through reconnaissance field work and laboratory studies to major field and laboratory study programmes. In most cases, the investigations undertaken for each step are similar, emphasis gradually shifting from the general to the specific. Therefore, in subsequent major sections of this report, the investigations are discussed by topic rather than by stage.

---

[5] The details of the investigations described in this section and the activities listed in Fig.1, being of common interest to site investigations in deep continental geological formations, are taken in part from a more comprehensive and basic companion report: *Site Investigations for Repositories for Solid Radioactive Wastes in Deep Continental Geological Formations*, IAEA Technical Report No. 215 (1982).

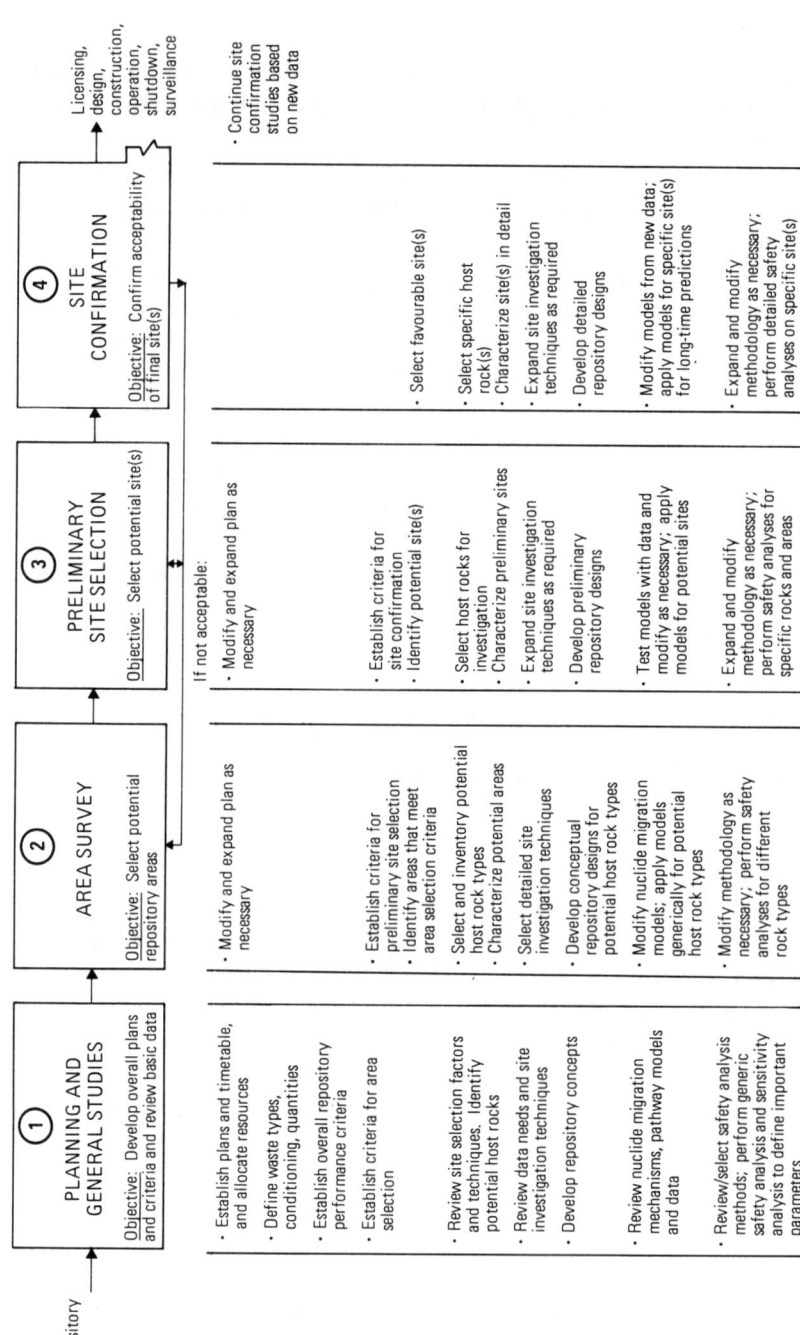

FIG.1. *Idealized sequence of investigations for site selection.*

Repository needs →

① PLANNING AND GENERAL STUDIES
Objective: Develop overall plans and criteria and review basic data

② AREA SURVEY
Objective: Select potential repository areas

③ PRELIMINARY SITE SELECTION
Objective: Select potential site(s)

④ SITE CONFIRMATION
Objective: Confirm acceptability of final site(s)

→ Licensing, design, construction, operation, shutdown, surveillance

If not acceptable:

① PLANNING AND GENERAL STUDIES
- Establish plans and timetable, and allocate resources
- Define waste types, conditioning, quantities
- Establish overall repository performance criteria
- Establish criteria for area selection

- Review site selection factors and techniques. Identify potential host rocks
- Review data needs and site investigation techniques
- Develop repository concepts

- Review nuclide migration mechanisms, pathway models and data

- Review/select safety analysis methods; perform generic safety analysis and sensitivity analysis to define important parameters

② AREA SURVEY
- Modify and expand plan as necessary

- Establish criteria for preliminary site selection
- Identify areas that meet area selection criteria
- Select and inventory potential host rock types
- Characterize potential areas
- Select detailed site investigation techniques
- Develop conceptual repository designs for potential host rock types

- Modify nuclide migration models; apply models generically for potential host rock types

- Modify methodology as necessary; perform safety analyses for different rock types

③ PRELIMINARY SITE SELECTION
- Modify and expand plan as necessary

- Establish criteria for site confirmation
- Identify potential site(s)
- Select host rocks for investigation
- Characterize preliminary sites
- Expand site investigation techniques as required
- Develop preliminary repository designs

- Test models with data and modify as necessary; apply models for potential sites

- Expand and modify methodology as necessary; perform safety analyses for specific rocks and areas

④ SITE CONFIRMATION
- Select favourable site(s)

- Select specific host rock(s)
- Characterize site(s) in detail
- Expand site investigation techniques as required
- Develop detailed repository designs

- Modify models from new data; apply models for specific site(s) for long-time predictions

- Expand and modify methodology as necessary; perform detailed safety analyses on specific site(s)

- Continue site confirmation studies based on new data

Note: At each stage of the site investigations, societal, ecological and national legislative issues are considered. The regulatory body should be involved according to national requirements. The term rock in this table includes all earthen materials including unconsolidated material such as soil (see Glossary).

14

5.1. PLANNING AND GENERAL STUDIES

The first step (Box 1 in Fig. 1) is to develop an overall site investigation plan, develop overall criteria and accumulate basic data. This planning receives input on the needs and timing for the repository and first develops an approach and timetable for the site selection investigations. From this overview, key decision points are defined and the activities are subdivided into more discrete investigations. The manpower, materials and equipment, time requirements and costs are estimated to the extent practicable, and responsibilities for the investigations are defined. Approval to proceed according to the plan and to provide the necessary resources must be obtained from the appropriate national authority. This plan is likely to require periodic updating, depending on the results of the investigations.

The types and quantities of wastes to be emplaced in the repository are defined and characterized. The likely final form and packaging of these wastes are identified. Overall performance criteria for the repository are developed, in conformity with national requirements. These criteria should take into account acceptable radiation dose commitments, longevity of repository protection of the wastes, and geoscience and engineering performance requirements for the repository. From these criteria, the geoscience criteria are established for selecting the general areas of interest for potential repository sites in Step 2.

Site selection factors are developed and used as a primary basis for identifying general types of geological materials that may be suitable to host a waste repository. Basic earth-science data are collected and sorted, from which the needs for new data are established. Information on available techniques for obtaining the new information is collected and sorted.

As part of this first step, the available methodologies for safety analyses are reviewed and the basic methods are selected. Available methods and models for analysis of radionuclide migration are reviewed and the requirements for model development identified. Of particular importance will be models describing groundwater movement and waste/groundwater/host-media interactions. From this point, a generic safety assessment is performed, including a sensitivity analysis so that the important parameters for future site investigations can be defined. Repository concepts are developed in parallel with and coordinated closely with these site investigation activities.

5.2. AREA SURVEYS

As the above studies approach completion, the area survey stage (Box 2 in Fig. 1) begins. Its objective is to select areas with favourable characteristics for a repository and to reduce them to a few preferred areas for further study. This is done largely by desk studies and some remote-sensing studies supported and evaluated by a small number of laboratory and reconnaissance field visits. Activities include, among others, mapping studies of geological relief, geomorphology, groundwater and surface-water systems. The data are evaluated and compared with criteria for selection of only a few preferred areas.

Based on information collected during the previous stage, potentially suitable host media are identified. Based on existing data and literature plus new investigations, a national inventory of areas worthy of more extensive field investigation is prepared. Such an inventory would include all host media potentially suitable for siting a waste repository. The geographic distribution and areal extent of such geological media, information on their homogeneity, depth and thickness, and their tectonic and hydrogeological settings are compiled and classified. This information will serve as a primary basis for the identification of preferable study areas which are then characterized. Many media types could potentially isolate radioactive wastes if located in an appropriate geological environment and if repository and waste-form designs are compatible with the host medium. Thus, factors other than host-medium occurrences, such as groundwater flow-path length, topography, existing land use, or transport problems may dominate the selection of favourable areas. The size of areas identified might range from a few to thousands of square kilometres. Selection of a waste repository based on earth-science aspects may also be made by first selecting appropriate general geological environments and then searching those environments for acceptable host media.

The work initiated within Step 1 on radionuclide transport and safety analyses and their related model development, investigation techniques and repository design is continued and directed towards studies on appropriate host media.

Because of the relatively large size of many of the areas selected, it is probable that an area will contain a number of sites that might be suitable for detailed investigation. At this stage, therefore, criteria for selection of more specific sites are established so that potential sites can be defined.

In smaller countries or in countries with extensive geological/hydrological information, it may be possible to identify specific potential sites more directly. It may further be possible to narrow the location process very quickly in some countries in which societal factors may dominate the selection process. On the other hand, there may be a need in larger countries or countries with complex geology to refine this selection in more than one step.

5.3. PRELIMINARY SITE SELECTION

When the previous stage is nearing completion, the preliminary site selection stage is initiated. The objective of this next stage is to select one or more potential sites for specific investigations which will provide detailed information and assessments on the characteristics of these potential sites (Box 3 in Fig.1). Investigative techniques are expanded as necessary for the more detailed studies. Earth-science studies are usually similar to those in the previous step but are undertaken in significantly more detail and on specific host rocks. They are undertaken through laboratory and field studies and include limited sampling and their associated data evaluations. The principal areas from which information will be gathered and assessed include: (a) earth-science aspects including geological, hydrogeological, geochemical and climatological data, and (b) waste capacity of the potential sites.

Radionuclide migration studies are carried out with site-specific data. The geometry of the body of the host medium is defined. The investigations will permit the collection of material for laboratory testing. Radionuclide migration and safety analysis methods and models are modified and formally applied to the specific conditions for the sites and host media under consideration. From these studies, criteria for the results of site confirmation studies are developed. Information is provided to allow the development of preliminary repository designs for the individual sites and evaluation of costs and logistics. Ecological and societal studies are also performed at a preliminary level.

Data are evaluated and compared with criteria and other selection factors in order to select a highly preferred site for which there is great confidence in its suitability as a repository.

5.4. SITE CONFIRMATION

As the best potential sites are selected, site confirmation studies are conducted (Box 4 in Fig.1). The objective of this stage is to confirm the validity of the results of earlier studies in greater detail and to confirm the suitability of the site(s). Earth-science studies here are expanded from those previously made and are concentrated on the site(s) of interest. Details of the site(s) and its (their) surroundings are defined through the use of field, laboratory and subsurface studies. From the results of this work, detailed specifications for the engineering characteristics of the site are established to allow detailed design of the repository.

Radiological transport and ecological evaluations are carried out in detail. Safety analysis models are upgraded for the specific site(s), and a detailed safety analysis is performed using all the detailed information available.

In-situ testing is conducted. Typical in-situ tests are as follows:

(a)   Hydrogeological: regional water flow and direction, near-field flow and direction, transport capability and capacity, and backfill and plugging effects; and

(b)   Geochemical: media/waste interactions, sorption/desorption processes, development of data collection techniques, and testing of chemical barriers to reduce radionuclide transport.

All the study results are brought together and summarized. Careful comparisons with criteria are made to ensure that the site will perform as required. Upon confirmation of the suitability of the site, appropriate authorities are provided with sufficient information to permit decisions to be made on authorization for design and construction of the facility.

During design, construction, operation, shut-down, sealing and surveillance of the facility, some confirmation studies may be continued as the activities provide additional specific information that will allow improvement of the validity of the earth-science studies.

A final site assessment based on all the investigations and evaluations may be prepared. Such an assessment would summarize all the relevant data, evaluations and conclusions derived from all site investigation, selection and confirmation activities.

6.   EARTH-SCIENCE INVESTIGATIVE NEEDS

This section identifies the major earth-science factors that should be considered for the investigations leading to the selection of a suitable site for a shallow-ground repository. The section also offers general guidance for these investigations and for selection. The basic information given in this and the next section has some features in common with other IAEA reports on site investigation for underground disposal of radioactive wastes. Some details are therefore repeated in the interests of complete coverage.

The primary mechanism for potential movement of radionuclides from most shallow-ground repositories is contact with and transport by water. When solid radioactive wastes are placed beneath the land surface or in embankments built up with earthen materials, they could be subject to dissolution and transport by percolating water in the partially saturated zone or by groundwater in the saturated zone. Radionuclides leached from the waste materials could then migrate downward and laterally away from the burial site or move upward by capillary action or by direct vapour transfer into the overlying soil zone, to be concentrated either by plants or evaporation to form salts at the surface. In nature, these transport processes are usually extremely slow and the subsurface flow paths of the fluids may be very tortuous. The principal mechanism that retards water-borne movement of radionuclides buried in the ground is sorption on soil and mineral particles. Site investigations therefore require information on the composition, permeability, porosity and sorption, and other properties of earthen materials around the site, as well as the properties affecting the water flow.

Other potential causes of initiating movement of radionuclides from a shallow-ground repository are processes that may expose the waste materials at the earth's surface. If this exposure should occur by such natural processes of erosion, the radionuclides could be transported by surface water or by the wind as volatile or particulate matter. Current knowledge of the geomorphic evolution and seismic activity in an area should enable selection of stable sites in order to minimize the potential of this type of radionuclide release from the repository, assuming that the design of the repository does not adversely affect this stability.

In both reconnaissance and detailed evaluations of a site, the time required for confining the radionuclides in order to prevent excessive harmful effects on man's environment must be a major consideration. Criteria of acceptability for a particular site partly depend on the nature of the radionuclides included in the waste.

After certain parameters that control the migration of waste constituents have been determined in the field, solute transport models may be used, recognizing the limits to their accuracy in predicting the time required for a radionuclide to move from the disposal site to man's environment, and in predicting its concentration level at various points. It must be realized that all the factors considered in evaluating the suitability of a site for disposal operations must be viewed in their entirety. A site with certain topographic and geological characteristics that might be excellent under a given climatic regime or existing pattern of water use in the area might be unsuitable if, for instance, the demographic distribution were unfavourable.

The establishment of a buffer zone around the repository where there would be restricted human activity can provide for additional safety of the repository during the period of institutional control.

The scientific investigative needs for repository site selection are discussed in the following subsections, as categorized by the general type of earth-science information involved. It is pointed out, however, that these various types of information are not entirely independent, and one frequently affects another in important ways.

6.1. CLIMATOLOGY

The climate of the area around a potential shallow-ground repository is an important factor in the consideration or selection of a site. The type and amount of precipitation will influence the amount of surface erosion, the occurrence and depth of groundwater, leaching and transport of radionuclides from waste materials, and the rate of evapotranspiration in the area. Wind directions and velocities should be considered because of their influence on erosion. Some examples of important climate parameters are discussed below.

Annual precipitation can range from negligible precipitation in arid regions to more than 1000 cm in very humid regions. Variations in precipitation with time are more noticeable in the semi-arid and arid regions than in the humid regions. In arid regions it is not uncommon for the annual rainfall to occur within a short period of a few days or a few weeks. Annual precipitation does not vary in a completely random manner, but neither is there a high degree of consistent precipitation. Long-term records of precipitation, when smoothed by various methods, have the appearance of distinct cyclic variations, but thus far none seems to fit any specific cycle. Precipitation records are useful, however, in providing general seasonal variations and frequencies of high-intensity rainfall.

Areas with humid climates generally lack a significant depth of unsaturated strata above the water table (i.e. shallow water tables), have short groundwater paths to surface water, and are prone to surface flooding, but they have relatively high evapotranspiration rates due to thick vegetation. Arid climates, on the other hand, may have deep water tables with extended subsurface flow paths to surface water, low frequency of flooding and very reduced evapotranspiration rates. Thus, arid climates are, in general, more favourable than humid climates for shallow-ground repositories.

Evaporation processes return a large part of annual precipitation to the atmosphere. Much of the surface water and near-surface soil moisture is evapotranspired at a rate varying with temperature, vegetation and the availability of moisture.

Evapotranspiration rates vary considerably with season and location. Summer evapotranspiration in northern latitudes ranges up to 80% or more of the annual total, whereas in subtropical latitudes the summer evaporation frequently constitutes less than 60% of the annual total. Transpiration rates depend on vegetation and growing season. Although other factors are important in evapo-transpiration, temperature has a major effect, with high temperatures contributing to high evapotranspiration rates.

The potential for increasing ice or snow accumulation and their possible erosional effects may be a factor in some areas.

6.2. TOPOGRAPHY

The topography of an area is a factor that must be considered in site investiga-tions and the operation of a proposed repository.

Topographic factors that affect the surface hydrology of a site are: size of the contributing drainage area, shape of the drainage area, gradient of the land surface, density of the drainage network, and slope of the major stream channels. Accurate determination of the effects of each of these factors on surface hydrology is nearly impossible because of interaction among factors and the virtual impossibility of separating the effect of one from another. However, some assessments can be made of all the factors combined.

Areas with topographic extremes could give rise to difficulties for a shallow-ground repository because they limit the types of engineered repository facilities that are applicable and increase the potential for erosion. Steep slopes limit the size, orientation and accessibility of disposal trenches or may even prevent trenches from being used. Such areas may also increase the potential for accidents in the transport of waste and may increase the cost of constructing and maintaining railways and roads to the repository.

If major earth-moving is to be carried out to modify the land surface for better accommodation of a repository, it is essential that due consideration be given to any changes in the rate of erosion or surface run-off that might ensue. In some cases, changing the topography may increase the local rate of water infiltration and result in a rise in the water table which could increase the possibility of the groundwater contacting the waste.

6.3. GEOLOGY

The safe disposal of radioactive waste requires that the geological environment of the area be favourable. The safety of the waste repository is highly dependent upon the engineered barriers and the characteristics of the natural geological formations that act as barriers to radionuclide transport. These natural barriers include the capability of the geological environment to minimize contact of groundwater with the waste, and its capability to sorb and retard the movement of waste constituents. Additional barriers may be provided to enhance the natural barriers by conditioning the waste and/or by application of other engineered features. These man-made barriers should be compatible with the natural barriers.

General hydrogeological factors to be considered in site investigations for shallow-ground waste repositories are given in Table III. It is, of course, not necessary for an acceptable site to include the most favourable types of all the factors. The suitability of a site will inevitably involve a balance between the more and the less desirable features. Each of the factors given in Table III embraces a broad field of characteristics, a brief discussion of which is given in the subsections that follow.

6.3.1. Geological structure and stratigraphy

The geological structure of the repository area and the stratigraphy of various geological formations, together with their spatial and age relationships, are

TABLE III. GENERAL GEOLOGICAL FACTORS
TO BE CONSIDERED IN INVESTIGATING
POTENTIAL REPOSITORY SITES

---

1. Geological structure and stratigraphy
2. Erosion characteristics
3. Lithology and mineralogy
4. Geotechnical properties of the rocks or soil
5. Tectonics and seismicity

---

fundamental to understanding the flow of water through the area. The water pathway, in turn, strongly controls the potential direction and rate of potential transport of radionuclides from emplaced wastes. Ideally, the area should have a relatively simple geological structure and uncomplicated stratigraphy so that the resultant hydrogeology can be defined with reasonable accuracy.

6.3.2. Erosion characteristics

Erosion includes all processes by which earthen materials are loosened and moved from place to place. Rates of erosion should be determined to establish that the waste will not be exposed by natural processes within the time-frame of the planned waste isolation. Two different conditions of erosion should be considered: (1) wind and water erosion that is typical of the area and is primarily dependent on the topographic and climatic conditions, and (2) changed erosion rates which could be caused by the implementation of the facility.

6.3.3. Lithology and mineralogy

The mineralogy of a geological stratum is important for underground repositories in that (a) it will dictate how the wastes may interact chemically with the formation rocks and soil and (b) it controls the chemistry of the ground-water. The ability of earth materials to sorb, or otherwise immobilize or retard, the movement of waterborne waste constituents is a principal reason why some environments are preferred for the disposal of radioactive wastes. Chemical reactions between waterborne radionuclides and minerals in soil and rock are influenced profoundly by the chemical characteristics of water. For these reasons, a knowledge is required of the groundwater chemistry at the repository site and its effects on possible chemical reactions between the radionuclides and local earth material and water.

22

6.3.4. Geotechnical properties

Soil and rock mechanics studies of the repository area will provide information on the physical and structural properties of the soil and bedrock materials to ensure that a particular repository design can be developed at the site. These studies provide physical property information not otherwise obtained by stratigraphic and petrographic observations. The data are needed for input to the design of the engineering aspects and man-made barriers of the disposal system.

It is desirable from the repository construction standpoint for the overburden to permit easy excavation of disposal trenches with conventional earth-moving equipment. Ideally, the texture of the overburden should be such as to permit the trenches to stand open with relatively steep sides without support. The depth of the overburden should preferably be thick enough to provide several metres of natural earth materials between the bottom of the disposal trench and the water table.

6.3.5. Tectonics and seismicity

Tectonic stability constitutes a favourable condition for a shallow-ground repository. A region that has been and/or is currently tectonically active may be characterized by contorted strata dislocated by a dense network of fractures and joints. The presence of a major fault frequently indicates a line of structural weakness and can be a principal pathway for groundwater to move, thereby reducing the capacity of the ground to confine the radioactive waste. Moreover, tectonically active areas frequently consist of complicated geological structures in which the movement of water can be difficult to predict.

Very often tectonic movements are discontinuous, giving rise to seismic events, e.g. earthquakes, where a gradual buildup of stress in geological strata is suddenly and catastrophically released as a movement of some areas relative to contiguous areas. In other cases very slow readjustment such as subsidence or uplift of the surface may be taking place and will continuously modify erosional or depositional processes. In areas of very active tectonics, uplift or subsidence of the ground could occur in the near future and modify the erosional processes and the surface and groundwater regimes. Recent tectonic deformation experienced by sedimentary formations, recent faulting, or modifications in the geodetic data can indicate general instability of the area in the past, present and future.

The seismic risk, or potential for earthquakes, is based on the potential frequency and intensity of earthquakes that have affected the area in the recent past. The seismic risk for a shallow-ground waste repository is of less concern than for siting a nuclear power station over the short term (i.e. a few tens of years) [5] but it has to be considered for a longer period of time. An earthquake could conceivably damage the engineered or natural structures of the repository, provoke collapse or failures of the confining barriers and thereby increase the potential rate of migration of the waste back to the surface. Earthquakes can also produce fissures and faults in the geological formations. Such features could possibly impair the site's capacity for confinement by increasing the permeability of rocks or by modifying the water-table elevation and characteristics of ground-water flow. Seismic activity can also trigger landslides and thereby modify topography, and can significantly modify the surface water and the groundwater regimes.

For these reasons, areas of high seismicity and those in the immediate vicinity of active faults should be considered unfavourable for choice as a shallow-ground repository.

6.4. HYDROLOGY

The principal natural means for the potential transfer of radioactivity away from the repository and into the environment is by water flow, either from surface water or from groundwater. Thus, a good knowledge of the characteristics of the water systems in the immediate vicinity and in the area around the repository is essential in order to assess the confining capacity of the site.

6.4.1. Surface hydrology

It is very important to understand the stream and lake networks in the vicinity of a repository site and to know their characteristics, such as flow rates and levels at different periods of time. This knowledge of the surface streams, lakes, ponds, swamps, etc., is necessary for many reasons, which usually include one or more of the following:

Evaluation of the flooding potential,
Evaluation of the potential of erosion and sediment transport and possible changes in river channels,
Identification of the recharge and discharge areas for the underlying aquifer(s),
Evaluation of its use as a resource to man.

Knowledge of the flows of surface water is needed to define the ground-water balance which, in turn, is needed to analyse the potential for the undesirable entrance of water into the repository. In addition, because surface waters constitute important monitoring points around the repository, knowledge of their systems is needed to properly evaluate the data from monitoring.

Surface waters are likely recipients of possible releases of radionuclides from waste repositories and they provide locations for dispersion of materials throughout the environment. They also present the possibility of dilution of radionuclides that might have been released from a repository. Such dilution could be an advantage or disadvantage, depending on specific site conditions and waste constituents.

For sites along a large lake or seashore (especially in flat areas), consideration should also be given to the potential for flooding as a result of tsunamis, seiches or tidal waves. The circulation of water flow currents may also need to be understood if waste constituents could reach an estuary or sea in relatively high concentrations.

6.4.2. Hydrogeology

In nature there are no known regions where the ground is devoid of liquid water.

In arid areas it is possible to find closed hydrogeological systems in which the hydrological balance is at steady state, i.e. the evapotranspiration balances the water supply so that there is no lateral flow of water and no outlet except towards the atmosphere. Such situations are generally advantageous for shallow-ground disposal. However, it may be difficult to predict climatic changes, and consequently the stability of the system, for the period of concern.

Most prospective waste-disposal areas will generally have one or more known aquifers that could possibly allow for transport of radionuclides towards the natural or artificial outlets of these aquifers. Even clays normally regarded as 'impermeable' have a finite measurable permeability, albeit low.

Consequently, wherever the repository is sited, the hydrogeological environment must be studied and understood in detail to permit prediction of the potential pathways of migration of radionuclides and their interaction with the groundwater and rock materials. For this reason the regional hydrogeology must be carefully considered at the early stages of site investigations. The required information related to underground hydrology for repository site investigations is summarized in Table IV.

The long-term stability of the conditions of the partially saturated zone and the groundwater must be considered for the period during which the waste remains a nuisance. This stability can be influenced by factors such as changes in the climatic conditions, changes in the regional base level of streams which can produce a modification of erosion rate, artificial modifications of the stream regimes resulting from channelling rivers, building dams, increased water abstraction, etc.

It is easier to predict the movement of groundwater for homogeneous strata with an intergranular porosity than for fractured rocks. Safety analysis and monitoring are consequently easier to perform and are more reliable in the

TABLE IV. HYDROGEOLOGICAL DATA NEEDED FOR REPOSITORY
SITE INVESTIGATIONS

1. Geological structure and stratigraphy.

2. Identification of lithologies with high and low permeability, their thickness, homogeneity and extent.

3. Water table or piezometric (water pressure) contours of all significant aquifers.

4. Well hydrographs (water flows) and maximum water table fluctuations.

5. Relationship between groundwater and surface water, the positions of groundwater and surface water use, positions of springs on land or under water, areas of recharge and drainage.

6. Flow in perennial and intermittent streams adjacent to the repository.

7. Ratio of precipitation to pan evaporation and the seasonal distribution of precipitation.

8. Detailed measurement of physical and chemical properties of aquifers in the laboratory, and physical properties measured by pumping or injection tests. Important aquifer properties include transmissivity, water content, effective and total porosity, dispersion coefficients, grain size and pore size distribution.

9. Relationship and degree of hydraulic continuity between aquifers.

10. Detailed hydrogeochemical and physical investigations of the unsaturated and saturated zones to determine migration rates of various solutes.

11. Seasonal variations in soil moisture tension.

12. Field and laboratory tests for radionuclide sorption in various lithologies.

homogeneous strata. Thus, from this standpoint, favourable strata are those with low permeability and which are homogeneous.

Favourable geohydrological conditions for a shallow-ground repository would, in general, include a homogeneous formation of low permeability, with an individualized and limited groundwater system, having no connection with other aquifers, and having a limited number of well-known outlets that are easy to monitor and control.

7. TECHNIQUES FOR EARTH-SCIENCE INVESTIGATIONS

The earth sciences which require particular study in the selection of a shallow-ground repository are outlined in Sections 5 and 6 and broadly include geology, hydrology and geomorphology. This section describes the techniques available within these earth sciences to obtain the data needed to evaluate a potential repository and to predict the behaviour of the radioactive wastes after disposal.

Investigative techniques to be used will depend on the stage of the investigations, which in turn controls the size of the areas being investigated. Initially, relatively large areas will be considered, from which favourable regions may be selected for more detailed study by applying broad criteria. Consequently, the techniques applied to large areas should ideally be capable of obtaining data relatively rapidly. The potential site areas being studied during the initial site investigation stages may vary in size from thousands to only hundreds of square kilometres.

It is desirable, for reasons of economy, to locate repositories near the area of major sources of wastes, i.e. near nuclear installations. Consequently, a large amount of data for potential repository areas may already have been collected in selecting sites for nuclear or perhaps other industrial installations.

The discussions in the remainder of this section are arbitrarily divided into the two categories of regional and detailed investigations. These generally apply to the site investigation stages indicated in Fig.1, in Boxes 1 and 2 and Boxes 3 and 4, respectively. However, a number of investigatory techniques can be used for general and for detailed studies, and thus there may in some cases be considerable flexibility about when some techniques are used.

7.1. TECHNIQUES FOR REGIONAL INVESTIGATIONS

Regional investigations are defined here as those that will yield appropriate data on areas for which it is impractical to undertake more than a relatively small amount of field study. Such techniques usually involve the identification of surface features as shown on maps and in reports which may by inference provide an insight into the subsurface. Also included are remote-sensing and airborne geophysics and limited field investigations which can give information on the subsurface geology. The techniques apply more generally to the investigative stages indicated in Boxes 1 and 2 of Fig.1, but can also be used to augment the information collected at later investigative stages.

7.1.1. Desk studies

The early site investigation stages will, to a large degree, involve the collection of data from all available published or unpublished sources. This will involve a review by professional earth scientists, in conjunction with waste management experts, of all available data on mineral and water resources, urban development, access facilities, land use, climate, topography, geology and hydrology. Using appropriate site selection criteria, the early studies should delineate areas that can be removed from further consideration for earth-science or other reasons so that the subsequent site investigations can concentrate on more detailed earth-science aspects. Available and published information is also extensively used, but to a lesser degree, for the later stages of site investigations on smaller site areas.

7.1.2. Remote sensing

Remote sensing of certain properties of the earth, using recording equipment on a satellite or on a low-flying aircraft, can provide some information relatively quickly and economically. Usually no one method can be relied on to identify all features, and therefore the results of a number of techniques are sometimes used together. All remote-sensing observations ultimately need verification by means of investigative techniques on and in the ground, but taken together they can provide a powerful tool for directing subsequent investigations. Remote-sensing techniques are particularly important in the reconnaissance of large unmapped areas where little or no previous work has been done. Even in areas where previous earth-based work has been done, remote sensing can be advantageous in verifying early interpretation of areas.

A number of techniques have been used to interpret geological and hydrological features from the air. Aerial photography has been used extensively in the past, while radar imagery and infra-red scanner imagery are recent additions. These techniques cannot penetrate beneath the surface and they rely on a surface manifestation on the underlying features being effective in elucidating the subsurface geology. However, airborne geophysical methods, such as measurement of the earth's magnetic or electromagnetic field, rely principally on the properties of the underlying structure and rock type. Remote sensing can often reveal features which, because of their scale, are not discernible by surface mapping. Features such as fault lineaments, fracture orientation, large-scale folds and variations in vegetation due to differing rock or soil compositions or moisture content, may be easily recognized from the air. Topographic features such as density of drainage may also give invaluable clues to the permeability of the underlying rocks since a large number of drainage channels are likely to be developed on relatively impermeable strata.

The best sensing methods will depend on the terrain and the features which are significant for interpretation. No one method is generally superior, but skill is needed in choosing the right methods for a particular area.

7.1.2.1. Aerial photography

Aerial photography is useful from the time of initial evaluation of an area for the development of detailed topographic maps. By using a number of points that have had previous surveys as guides, coordination can be effected with areas that have already been aerially photographed, and thus minimize the need for new surveys.

Simple monochromatic photographs, in addition to identifying obvious details such as surface features, surface drainage patterns, vegetative cover and topography, can yield information on large-scale geological structures such as folds, faults and fracture directions which, because of their scale, cannot easily be recognized by surface mapping. The experienced scientist can infer or interpret a great deal of information on the physical and chemical properties of the rock types from aerial photography; for example, the density of drainage features is directly related to the permeability of the rock, while the identification of vegetation can indicate the rock types and may be used to map lithological boundaries.

Monochromatic aerial photographs are the basic tool for exploration and field mapping. They are relatively inexpensive to acquire and are readily available for many areas. Aerial photographs are usually obtained when the sun is high so that the effects of shadow can be minimized. However, for geological mapping, low-sun-angle photography may be useful to identify subtle differences in relief or textural patterns, especially in areas of low relief. These are obtained when the sun angle is $10°$ or less above the horizon and should be used in addition to the high-angle photographs.

Colour and infra-red (IR) colour photographs are superior to monochrome in that the eye can distinguish more shades of colour than of grey, so that colour differences attributed to different rocks, soils, mineralization or vegetation may be more easily recognized. IR colour film is often superior to colour or monochrome film for geological exploration because vegetation has a much wider range of reflectance in the IR region than in the green regions. IR may be used to discriminate more effectively between various species and growth stages and may be used to identify soil and rock type, moisture content and possibly trace-element distribution. Water appears blue or black on IR colour film and is readily distinguishable from vegetation, which generally appears red.

Multispectral photography, in which a wide range of discrete radiant spectral bands are recorded simultaneously, may in some cases facilitate the recognition of a particular feature. This technique is still in its experimental stages but may prove useful for geological interpretation in the future.

7.1.2.2. Radar imagery

This involves transmitting pulses of microwaves and receiving the back-scattered energy. It is particularly useful for surface structural mapping, especially as data can be gathered in poor weather. Radar has some unique capabilities which, although not necessarily superior to other techniques per se, are useful for reconnaissance work. The main advantages of radar imagery lie in its capability during one scan to produce good images several hundred kilometres long by tens of kilometres wide which cannot be obtained at this scale by aerial photography. Oblique 'illumination' and viewing by radar is especially good for highlighting faults and fractures; a further advantage lies in its suppression of minor details such as vegetation, thus producing an uncluttered image from which faint lineaments such as faults can be more easily recognized.

7.1.2.3. Infra-red scanner imagery

Certain wavelengths in the IR range radiated by rocks to a degree dependent on their temperature and emissivity can be sensed remotely. IR imagery can therefore be used to differentiate rocks according to their thermal properties, basically by identifying hot and cold areas on the ground. For geological purposes it is essential to obtain data at night in order to avoid interference by solar radiation, and the correct weather conditions are particularly important for effective imagery. For this reason, surface data such as air temperature, wind speed and direction, and relative humidity should be collected on the surface to aid interpretation. Apart from identifying volcanic features or areas with potential gas or thermal reserves, IR imagery has been most successful in arid and semi-arid areas where there is little interference from vegetation and where surface temperatures are related largely to moisture content. Scanning large bodies of water where the surface is regular also allows for more accurate temperature assessment and may be used to investigate mixing of water with thermal contrasts.

7.1.2.4. Airborne geophysics

Aeromagnetic surveys in which the magnitude of the earth's local vertical magnetic field is measured form the basic geophysical method for regional studies. They were mainly developed for detecting magnetic ore bodies but have many geological applications such as determining the distribution of igneous intrusions, major faults or the depth of sedimentary rocks overlying the basement rocks.

Electromagnetic surveys which measure either natural fields or induced electromagnetic fields in the earth are used principally for mineral exploration. Induced electromagnetic fields in which a transmitter coil and receiver are mounted on an aircraft are susceptible to the effects of a conductive overburden which limit the depth of penetration. In this way near-surface mineral or clay deposits may be identified and the technique may also be used to distinguish shallow mineralized groundwater in unconfined aquifers.

Natural electromagnetic surveys based on the field response to natural discharges of lightning can also be used. In this case the lower frequencies involved allow for penetration 200 to 2000 metres below the earth's surface. The technique has not been developed owing to reliance on natural electric storms, and there are some problems about interpretation, particularly since a number of atmospheric discharges may occur simultaneously.

Very low frequency (VLF) radio field methods, which use plane waves from communication radio stations sometimes 4800 km from the area of interest, have been found to be sensitive to large-scale geological structure to a depth of 200 m. While aeromagnetic maps are related to the rock's magnetic susceptibility, radio methods which also measure the magnetic field component are related to conductivity lineaments which are affected by foliation, compression and shearing as opposed to lithological layering. The horizontal electric field associated with these transmissions can also be measured to determine the electrical resistivity of the formation and produce maps similar to surface resistivity surveys. Consequently, aero-resistivity may be used to identify resistive formations such as unsaturated sand and gravel reserves or areas of permafrost, or conductive formations (e.g. clay deposits), mineralized groundwater or geothermal areas.

For low-altitude sensing, gamma-ray spectrometry can be useful in determining the presence of radioactive minerals which can be used to interpret surface lithology to a depth of 0.3 m or less.

7.2. TECHNIQUES FOR DETAILED SITE INVESTIGATIONS

After potentially favourable regions have been identified, more detailed site investigations can be carried out to define the geology, hydrology and prevailing geomorphological processes. The techniques discussed below apply generally to the investigative stages indicated in Boxes 3 and 4 of Fig.1, but some can also be applied, to a lesser extent, to earlier investigative stages.

A detailed knowledge of the subsurface geology and hydrogeology is required, to be able ultimately to predict radionuclide movements, and this can only be achieved satisfactorily by means of boreholes. However, a number of surface techniques that are usually relatively inexpensive and rapid can be used to gain an insight into the underlying geology and hydrogeology and may be used to advantage in locating boreholes. In general, a combination of boreholes and surface techniques appropriate to the geological or hydrogeological environment will provide the optimum cost-benefit ratio.

Surface techniques include geological mapping, surface geophysics and geochemical sampling from which information on the subsurface geology and hydrogeology may be inferred. Geomorphology and surface hydrology relate specifically to surface features and processes, while hydrogeological investigations rely heavily on information gained from the analysis of rock cores obtained through boreholes to obtain information on the in-situ hydraulic properties of the geological formations. Borehole geophysics is an important tool for hydrogeological investigations and is an aid in geological interpretation. Geotechnical investigations may be carried out on rocks at the surface or may involve laboratory testing of rock samples from boreholes.

7.2.1. Drilling techniques

There are many ways of drilling boreholes, depending on the geological formations, the depth required and the purpose of the investigation.

(a) *Auger methods:* Two types of augers are common: the continuous flight for holes up to 250 mm diameter and bucket augers for wells greater than 1 m diameter. Both types of augers are rotated into the ground either by hand or by mechanical means and provide disturbed samples in soft or unconsolidated formations. Hollow-stem augers may also be used to provide less disturbed core samples.

(b) *Percussion methods:* These techniques involve repeatedly lifting and dropping a steel tube into the ground; this action loosens unconsolidated formations or crushes consolidated rocks. The loosened material may be removed from the hole by a bailer or, less commonly, a sand pump. With this method. smaller-diameter tubes, 0.5 m long, fitted with a cutting shoe, can be driven carefully into unconsolidated formations to obtain relatively undisturbed samples from known depths.

(c) *Rotary drilling:* Several variations of this method are possible. The technique consists of rotating a hollow drill rod fitted with a cutting bit while water, mud, air or foam is pumped down the drill stem to return the rock cuttings to the surface on the outside of the drill stem. Reverse rotary drilling is used where the fluid, usually mud or water, is pumped down the annulus and returned up the drill stem. This technique is used mainly for large-diameter holes in unconsolidated formations. Rotary drilling can be used to recover rock cores in medium to well consolidated formations using an appropriate coring bit. The resultant cores are relatively undisturbed.

All drilling methods result in at least some disturbance of recovered rock samples and the geological formation immediately round the borehole. When testing in boreholes either by pumping or geophysical methods (described later), it is important to take this factor into account. Shallow trenches made with a hydraulic excavator can also provide considerable information on the geology, especially for lithological characterization such as the orientation of bedding and concentration of fractures.

Whenever boreholes are drilled it is highly desirable to log accurately the different strata penetrated and the occurrence of groundwater. Moreover, rock samples should be taken either as cores or in the disturbed form. Specialized sampling and preservation techniques are required for satisfactory water sample collection so that changes in the natural chemistry do not occur before laboratory analysis can be carried out.

7.2.2. Geological mapping

Surface geological mapping is fundamental to understanding the geological structure of the area and is the basic technique of a geologist. It is a study of the distribution of rock types, their orientation, and their representation on a topographic map. Any desired geological feature may be represented on a map; common geological maps display only the bedrock geology, or may show both the bedrock and overlying superficial deposits (solid and drift maps). More specialized maps may display orientation of fractures, cleavage or tectonic trends as well as fault patterns, while other maps may elaborate on the distribution and thickness of a particular mineral resource.

7.2.2.1. Surface geological mapping

Surface mapping requires sufficient exposure of rocks on the surface to allow the areal, spatial and temporal relationships among rock types to be distinguished. Structural features such as faults or folds can be mapped or inferred, and features such as fractures, cleavage and the characteristics of contacts between the rock strata are used together to build up the tectonic history of the area.

The degree of detail given in a geological map depends on the scale of the map being prepared and the available rock exposure. In general, a map on a scale in the order of 1:10 000 is required at the start of a detailed site survey, but a smaller-scale map will be required as the investigation proceeds. Rock exposure at the surface can vary greatly depending on the amount of vegetative or soil cover. Ambiguities can result through insufficient data. Surface mapping therefore uses all available information that may relate to the geology, e.g. the type and distribution of vegetation or soil types, or changes in topography, that may indicate the underlying lithology, but nevertheless at this stage boreholes or shallow excavations may be necessary to provide satisfactory data.

7.2.2.2. Subsurface geological mapping

Subsurface geological mapping is undertaken as the detailed site investigations proceed. Maps on a scale 1:10 000 and smaller are required in the detailed investigations.

Bedrock geological maps are developed to indicate the boundaries and extent of the formations beneath the superficial deposits. Such maps are commonly accompanied by one or more cross-sections which show the vertical relationship of the mapped formations.

Bedrock maps often carry enough information to provide an understanding of the structural geology of the area. In areas with highly complex geology, however, where a large amount of structural data is necessary to interpret the subsurface geology, special maps are prepared to show formations with similar structure. These maps show relatively detailed structural characteristics and can have many symbols to represent characteristics of folds, faults, joints and different kinds of sedimentary and flow structures. Such maps may also show structural contours and isopachs. Structural contours are drawn on top of those for a particular formation or stratum; isopach lines, which connect points where the formations have equal thickness, are drawn. Structural geological maps are particularly useful for determining the occurrence of groundwater and conditions for underground installations.

Maps can be produced to show many desired parameters, such as: hydrogeological maps showing distribution of aquifers and variations in water quality; geotechnical and soil maps showing the extent of superficial formations and their engineering characteristics; and resource maps showing the extent of mineral resources.

7.2.3. Near-surface geophysical techniques

These techniques are useful aids in most geological surveys and in some cases can be used for hydrogeological interpretation. Geophysical surveys may range from very simple measurements requiring one or two people to large-scale field operations requiring sophisticated equipment and vehicles, and upwards of 100 people. Similarly, data processing and interpretation may range from plotting a simple graph to the use of digital computors. In the latter case the processing costs may be large and comparable to the cost of data acquisition in the field.

All geophysical methods require measurements of certain physical properties of the earth and in many cases require control by drilling boreholes.

7.2.3.1. Electrical methods

Electric prospecting is based on the fact that rocks conduct electricity to differing degrees depending on their mineralogy, porosity, degree of saturation and conductivity of pore fluids, and the distribution of fractures or orientation of minerals. Conductivity in a rock is due to electrons moving either through the rock matrix or (for the greater part) to the movement of ions through the interstitial fluids. The conductivity (or its converse, resistivity) of rocks may be studied by measuring the distribution of electrical potential on the surface as a result of natural electric fields or by artificially induced electric fields. Electrical methods can provide information on stratigraphy, geological structure and faulting, and on the occurrence and quality of groundwater.

34

7.2.3.1.1. Telluric current

This method involves the measurement of horizontal components of electric currents circulating within the earth's crust. Regional perturbations of the telluric field are related to the heterogeneity of the ground. The difference of the potential at the ground surface is a direct ratio of the resistance of the underlying geological formation. The method may be used to determine the location of deep bedrock and to identify the structure of large sedimentary basins.

7.2.3.1.2. Magneto-telluric

This method consists in determining the vertical variation of the apparent electrical resistivity of the earthen materials. Measurements of spontaneous potentials on the earth's surface vary with the period (1/frequency) of the magneto-telluric waves and are related to the electrochemical activity that would occur between one ore body in contact with fluids of differing compositions. Potentials may also be set up by the flow of water through a porous medium (streaming potentials).

7.2.3.1.3. Resistivity

This method is based on variations in electrical resistivity in the ground due to changes in lithology, water content and mineralogical properties. Because it is easy to use and portable, it is extensively used for hydrogeological investigations. The most common method is to measure differences in electrical potential induced by applying an external electric current between two electrodes. The distance between the current electrodes determines the depth of the formation under investigation. If a correlation of resistivity with depth can be made with a hydrological log from a nearby borehole, the technique can be used to interpret changes in resistivity in surrounding areas.

In relatively homogeneous and porous formations, resistivity depends on water content and water conductivity. Thus, resistivity can provide a guide to the depth to groundwater. It may also be used to define the distribution of pollution plumes (usually of high conductivity owing to dissolved minerals) around sanitary landfills, and in this application the direction of groundwater flow can also be inferred. Other applications include mapping the distribution of buried channels, detecting the presence of a conducting layer between two aquifers, defining fault zones, mapping the position of saline water interfaces, and, in some favourable cases, determining the depth to the water table.

7.2.3.1.4. Spontaneous potential

This method involves the measurement of natural electrical potentials existing on the ground surface. These can be related to the thickness and depth of conducting formations and therefore to the presence of geological features such as faults, igneous intrusions, ore bodies or voids (in karstic terrains).

7.2.3.1.5. Drop-of-potential-ratios

This method, derived from the spontaneous potential method, consists in measuring the potential difference between a fixed electrode and a series of electrodes placed in the ground at varying linear distances. The method can be useful for indicating the position of a fault, void, ore body or intrusion.

7.2.3.1.6. Induced polarization methods

These methods, which rely on a study of the residual potential after the current between two electrodes has stopped, are also used for detecting ore bodies although the exact physicochemical mechanisms that produce the effect are not well understood. The system has been used for differentiating horizons by their differing responses to polarizing potentials where the resistivity method cannot be used.

7.2.3.1.7. Electromagnetic method

This consists in measuring natural or induced magnetic fields and is generally used to locate metallic ore bodies which generate a magnetic field when subjected to an alternating electric current. The method may also be used to indicate the presence of faults and fractures but has little application to direct hydrogeological studies.

7.2.3.2. Seismic methods

Seismic prospecting has been extensively developed in the search for oil and other minerals. Although the propagation of seismic waves depends principally on the density, porosity, extent of fissuring, water content and degree of weathering in rocks, the technique is mainly used to elucidate geological structure by interpreting the time taken for seismic waves produced at a point to be refracted or reflected within the earth and return to a line of detectors (geophones or seismometers) usually placed in a line some distance away on the surface. The velocity of the seismic waves is low through unconsolidated and detrital rocks and is higher in compact and dense formations. Seismic velocity generally increases with depth.

The seismic source can consist of an explosive charge placed either on the surface, in shallow hand-dug excavations, or in tamped boreholes. Alternatively, where shallow investigations are being undertaken, a falling weight or sledge-hammer may be enough to create a sufficient disturbance. Non-explosive sources such as vibrating pistons have also been successfully developed for reflection surveys with little or no damage to the terrain.

The investigation of *reflected wave forms* is the predominant seismic method in use, especially for identifying the geological structure of oilfields or mineral deposits. The method relies on the interpretation of seismic signals reflected from the interface between lithologies with contrasting seismic velocities, e.g. limestone and shale. The technique requires knowledge of (a) surface topography, (b) the thickness and seismic velocity of the surface-weathered zone, and (c) the vertical seismic velocity profile obtained from deep boreholes in the strata under investigation. However, it often cannot be used in shallow investigations because of interference by refracted waves produced by the low-velocity layer of weathered rock or overburden near the surface. The method is appropriate for definition of relatively deep formations.

Consequently, *seismic refraction* methods, which are, in comparison, much simpler and do not involve a large number of field personnel, sophisticated recording and processing equipment, or large seismic sorties for shallow work, are appropriate for near-surface geological or hydrogeological studies.

Seismic refraction measures the time taken for refracted seismic waves to pass through a formation and can be used for both deep (depth to basement rocks) and shallow investigations. The deeper the information required the larger is the seismic source required. Deep refraction observations are appropriate for basement-rock locations in large sedimentary basins, and it is common for the seismic source to consist of several hundred kilograms of dynamite for satisfactory energy transmission.

Shallow refraction studies are appropriate for detailed site investigations for shallow-ground repositories. The technique can be used to identify depth to bedrock, depth of the surface zone with low seismic velocity, the distribution of buried channels and, in cases where there is a thick saturated aquifer above bedrock to the depth of the water table, it can be used to detect faults or fault zones. Its use in defining hydrogeological regimes is questionable, but claims have been made that in favourable circumstances the method can be used to detect the interface between saturated and unsaturated zones.

7.2.3.2.1. Gravimetric surveys

These surveys give information on the differences in density among various kinds of geological formations. The technique consists in measuring the areal variation in gravity and provides information (after appropriate corrections) on the distribution of rock masses of different density or tectonic features, usually on a large scale. This information is mainly qualitative. For quantitative interpretation, complementary studies by drilling and sampling are needed; for example, density measurements have to be compared with measurements on reference samples of rocks.

Gravimetric surveys are best carried out on the ground. Aerial (airborne) gravimetry does not provide information useful for detailed geological or hydrogeological evaluations, but can be used in conjunction with other parameters for confirmation. These surveys are useful in identifying buried channels or changes at bedrock depth, especially where associated with faulting.

7.2.3.2.2. Magnetic surveys

Magnetic prospecting provides information on the differences in the magnetic characteristics of rocks, and this is mainly related to their content of ferromagnetic minerals such as magnetite and ilmenite. The magnetic field measurements (the vertical component and the total field measurement) are made with a magnometer and then modified with different corrections.

Airborne prospecting can be used for general qualitative reconnaissance. Land prospecting is used for detailed and more quantitative studies, usually in conjunction with gravimetric prospecting.

Like gravimetry, magnetic prospecting provides information on the general distribution of rock masses and tectonic features, e.g. faulting, but it may also indicate orientation of fracturing. Neither technique is appropriate for detailed hydrogeological studies, but together they may provide geological information from which the general hydrogeology can be inferred.

Unequivocal interpretation of magnetic anomalies may be difficult to achieve since magnetization of most rock formations is complex. The method is mainly used for identifying the distribution of igneous and metamorphic rocks between sediments which are non-magnetic except for magnetic rich sands and gravels. Metamorphic rocks often produce strong magnetic lineaments which can be correlated with cleavage, fracturing or faulting.

7.2.3.2.3. Surface radar

Ground-probing radar is a relatively new geophysical technique which can provide high-resolution data on surficial geology and groundwater distribution. Subsurface radar reflections are generated when abrupt changes in electrical properties occur in the ground owing to changes in material and soil water content.

In field tests, water table depths have been mapped and bedrock was delineated to depths exceeding 20 m. A great deal of graphic information can be extracted quickly and inexpensively by the surface radar method [6].

7.2.4. Borehole geophysical techniques

Borehole geophysical techniques have been developed largely for oil-well drilling to great depths where the cost of obtaining rock cores for laboratory study is high. In a detailed site investigation for a shallow-ground repository, it is envisaged that continuous samples or cores could be obtained for laboratory tests. Some geophysical logs used for identifying rock types would therefore be redundant to other tests described earlier. However, certain geophysical techniques, especially nuclear methods, dipmeter and television viewing, offer unique tools for hydrogeological and geological investigations.

Borehole geophysics can provide very useful continuous information about the characteristics of rocks and their fluid content, lithology, geometry, electrical resistivity, bulk density, porosity, permeability, moisture content, specific yield of water-bearing rocks, chemical composition of groundwater, etc. A large number of different types of geophysical well logs are available (a log can be defined as a sequential record of a particular parameter versus depth). The various types of geophysical and related logs are listed in Table V, and the most common types are described in the remainder of this subsection.

7.2.4.1. Electric logging

These logs are carried out in an uncased hole filled with water or mud; they rely on measurements of the natural electrochemical or induced electrical response of the formation. Specific techniques include:

Spontaneous potential log (SP): a graphic plot of the small differences in voltage that develop at the contacts or gaps between lithological units and the borehole fluid; used primarily to identify lithology and for stratigraphic correlation.

Resistivity log: a multi-electrode probe in which an electric current passes between two electrodes via the surrounding rock, the borehole fluids and the formation fluids in the rocks. Several modifications of the method and probe have been made giving the laterolog, microlaterolog, etc. The method is mainly used for correlation of the lithology with depth and to infer water content, porosity and salt content of water.

Dipmeter: in effect an electric log with four electrodes spaced radially on a probe. By measuring the spontaneous potential between the electrodes it is possible to infer the dip of the geological strata or orientation of fracturing or cleavage.

Single-point potential: similar to the spontaneous potential log except that one electrode is coupled to the potential at the ground surface. The method is used (a) for stratigraphic correlation, (b) to determine fractures, and (c) for confirming steel casing depths.

TABLE V. SUMMARY OF LOG APPLICATIONS FOR REPOSITORY SITE INVESTIGATIONS

Required information on the properties of rocks, fluid, wells, or the groundwater system	Widely available logging techniques which might be used
Lithology and stratigraphic correlation of aquifers and associated rocks	Electric, sonic or caliper logs made in open holes; nuclear logs made in open or cased holes
Total porosity or bulk density	Calibrated sonic logs in open holes; calibrated neutron or gamma-gamma logs in open or cased holes
Effective porosity or true resistivity	Calibrated log-normal resistivity logs
Clay or shale content	Gamma logs
Permeability	No direct measurement by logging; may be related to porosity, injectivity, sonic amplitude
Secondary permeability — fractures, solution openings	Caliper, sonic or borehole televiewer or television logs
Specific yield of unconfined aquifers	Calibrated neutron logs
Grain size	Possible relation to formation factor derived from electric logs
Location of water level or saturated zones	Electric, temperature or fluid conductivity in open hole or inside casing; neutron or gamma-gamma logs in open hole or outside casing
Moisture content	Calibrated neutron logs
Infiltration	Time-interval neutron logs under special circumstances or radioactive tracers

TABLE V. (cont.)

Required information on the properties of rocks, fluid, wells, or the groundwater system	Widely available logging techniques which might be used
Direction, velocity and path of groundwater flow	Single-well tracer techniques — point dilution and single-well pulse; multiwell tracer techniques
Dispersion, dilution and movement of waste	Fluid conductivity and temperature logs; gamma logs for some radioactive wastes; fluid sampler
Source and movement of water in a well	Injectivity profile; flowmeter or tracer logging during pumping or injection; temperature logs
Chemical and physical characteristics of water, including salinity, temperature, density and viscosity	Calibrated fluid conductivity and temperature in the well; neutron chloride logging outside casing; multi-electrode resistivity
Determining construction of existing wells, diameter and position of casing, perforations and screens	Gamma-gamma, caliper, collar and perforation locator, borehole television
Guide to screen setting	All logs providing data on the lithology, water-bearing characteristics, correlation and thickness of aquifers
Cementing	Caliper, temperature, gamma-gamma; acoustic for cement bond
Casing corrosion	Under some conditions, caliper or collar locator
Casing leaks and/or plugged screen	Tracer and flowmeter

(Data from Ref. [7] and Chapter E1 of Techniques of Water Resources Investigations, US Geological Survey.)

Induction log: gives the electrical conductivity of a formation by measuring the secondary magnetic field within the rock generated by an alternating magnetic field. Results can be used to infer porosity and fractures in dry boreholes.

7.2.4.2. Nuclear logging

Nuclear methods measure the natural or induced radioactivity of the formations penetrated by the borehole. The advantage of these techniques is that they can be used in either cased or open holes filled with any type of fluid. The most common logs are the following:

Natural gamma log: records the natural gamma-radiation emitters in rocks. It is used for identification of lithology and correlation of stratigraphy. It can indicate radioactive minerals usually concentrated in clays or shales.
Gamma-gamma log: records the intensity of induced gamma-radiations (emitted by an artificial source and backscattered after attenuation in the surrounding formation). It is mainly used to identify lithology and to measure the bulk density and porosity of rocks.
Neutron log: measures the thermal neutron flux or induced gamma-radiation flux resulting from a neutron-emitting source in a probe. The reduction of neutron velocity in the formation depends on the amount of hydrogen present or on the water content. A neutron log provides information on moisture content above the water table and on rock porosity below the water table.
Neutron-gamma log: measures gamma-radiation produced by irradiating the formation with neutrons. It can give porosity values in an uncased hole.
Spectral gamma log: measures the natural gamma-radiation over a particular spectrum. It can be used to determine lithological and mineralogical variations, especially radioactive K, U and Th, and the interrelation of different fissure systems.
Nuclear magnetic resonance log: can provide information on effective porosity and permeability but is not generally proven.

7.2.4.3. Acoustic or sonic logging

Acoustic logging measures the velocity of sound through the rocks. It is the record of the time between transmitting and receiving an acoustic pulse in a probe. The acoustic velocity in porous media depends on many factors, such as the nature and density of the rock matrix, the pore size distribution and porosity, properties of the interstitial fluids, etc. Acoustic logging is particularly useful in consolidated formations for stratigraphic correlations, for identifying fractures, and for measuring porosity. It can also be used to check the continuity of cement grout behind casing. Acoustic logging with a sound source (e.g. dynamite) at the surface and a geophone mechanically coupled to the borehole wall is necessary to relate seismic reflection surveys to stratigraphic logs.

Other sonic logs include the seisviewer, which in effect produces a television picture of the borehole wall using high-frequency sonic pulses that can give information on fractures and bedding planes.

7.2.4.4. Miscellaneous logs

Other kinds of logs available are described briefly as follows:

Television cameras can be lowered down a borehole to provide a continuous picture of the walls of the hole. They cannot be used in mud-filled or cased holes.

Caliper logs simply measure the diameter of the hole. They are used to correct other logs which are sensitive to pole diameter. The caliper log may identify fractures and, to a certain extent, lithology since the borehole diameter will be larger where rock fragments have broken away from the borehole in heavily fractured formations.

Temperature conductivity probes measure temperature and thermal conductivity in the borehole fluid. They are used for correcting other measured characteristics that are sensitive to these parameters. They are also useful in themselves to indicate inflow or outflow of water in a borehole and may also be used to infer the quality of the water and hence the depths at which water samples should be obtained in geochemical studies.

Flowmeter probes measure the vertical velocity of water flow in a borehole. They can use either an impeller, a heat pulse or a tracer. The water flow measurements are important for recognizing whether water obtained at a particular depth in a borehole is representative of that point in the aquifer and in studying the pressure differences between two aquifers. It can also be used to determine the proportion of water obtained from various parts of an aquifer when the borehole is being pumped.

Geochemical logs involve a variety of chemical-type probes, although to date most of them do not perform satisfactorily at depth. Dissolved oxygen probes are an exception and can indicate the presence of organic pollution in groundwater.

7.2.5. Geochemical techniques

Geochemical exploration methods commonly used to identify chemical anomalies which may correlate with economic ore deposits can assist somewhat in radionuclide migration studies; they may also be useful as an aid to geological mapping or to confirm that an area is devoid of complicating features. Such techniques involve the collection of soil and rock samples, stream and stream-bed samples and their chemical analysis.

Since laboratory conditions often do not simulate field conditions precisely, it is usually necessary to carry out field studies into rock-waste interactions and waste migration rates.

In the unsaturated zone that normally underlies a repository, chemical interactions under controlled conditions can be studied using lysimeters. These are in-situ isolated blocks of soil or rock, usually underlain by a floor of material with low permeability, in which water infiltration and drainage can be measured. Lysimeters may be made from tanks repacked with soil or unconsolidated rock embedded with their surfaces at ground level and fitted with horizontal drains above their base so that the amount of percolation may be measured. Alternatively, undisturbed blocks of rock may be isolated by choosing a place where the rock is underlain by a relatively impermeable clay horizon, digging trenches into the clay, and backfilling them with an impermeable material. Drainage ports at the bottom of the lysimeter, and sampling tubes at various depths below the surface, are led into an adjacent trench. The migration, sorption and dissolution of particular wastes introduced at the surface of the lysimeter can thus be studied. Destructive sampling at various times may also provide samples of contaminated rock for detailed geochemical analysis in the laboratory or for measuring water infiltration through the surface.

A useful intermediate method of investigating leachate interaction with unconsolidated materials involves the use of monolith lysimeters. These consist of glass-reinforced plastic tubes 1.2 m long and 0.8 m in diameter, which can be hydraulically driven into the formation. The relatively undisturbed material can thus be removed to the laboratory for instrumentation and irrigation with the expected leachate composition. Monoliths may be operated under unsaturated or saturated conditions to represent the water movement expected.

Various forms of instrumentation to measure soil-moisture content, soil suction, temperature and gas-phase composition are usually installed in lysimeters so that when they are coupled with infiltration and drainage rates they may be used to determine the hydraulic parameters for unsaturated flow.

The type of vegetation in an area may be recorded (geobotany), or particular plants may be chemically analysed for their uptake of particular elements (biogeochemistry). Vapour or gas surveys have similarly been used to detect gases such as H_2S, Rn, SO_2, CH_4 and isotopes which may be diffusing upwards from a buried mineral or ore deposit or from otherwise undetected geothermal activity.

7.2.6. Geomorphological investigations

From the existing geomorphology of a site, an understanding of the previous geological events of an area can be derived as an aid in predicting the events likely to follow. The subject can be approached in three ways:

(1) A study of the distribution of land forms.

(2) A study of the physical processes that operate in shaping the earth's surface.

(3) A study of surface deposits which indicate the chronology of past events.

In siting a shallow-ground repository, the physical processes which result in erosion of the earth's surface are usually the most important, although the other studies must not be overlooked. Detailed geomorphological mapping, for instance, may identify areas where landslides have occurred in the past, and slope analysis or the recognition of stream erosion patterns may indicate areas where such events may occur in the future and their consequential effect on the repository.

Geomorphological processes often require instrumentation for their study. The studies are often more time-consuming than the study of land forms because of the slow rate at which the processes take place. This is especially true if erosion processes are discontinuous, for example, if significant erosion is associated with flooding, which may occur only once in 20 years or more. The rate of erosion of an area depends on numerous climatic, geological and hydrological factors which must be considered together. It is often difficult to measure certain erosional effects since they may proceed slowly. If erosional effects are rapid enough to be measured, then the site may well be unsuitable as a repository.

Erosionally active zones are frequently associated with rivers. In any location, possible re-adjustment of an existing river channel, especially after flooding, should be considered.

Although a site may appear erosionally inactive locally, it should not be assumed that a present area of sedimentation will remain inactive. Apparently unconnected events (e.g. landslides or dam failures) may considerably affect drainage patterns and profoundly affect the erosional stability of an area. Isostatic re-adjustment following the melting of glacial ice packs can result in gradual uplift of an area, and changes in sea or lake levels can alter river profiles, leading to increased erosion or sedimentation.

Like geological investigation, geomorphology relies very much on a good topographical mapping as a basis for further studies. More detailed topographic measurements may be required in certain instances, especially where slope changes may provide remnants of a former erosion surface or where river-channel elevations are required to identify channel regrading as a result of changes in level of a lake or sea into which the river drains.

Aerial photographs can be extremely useful in this respect since large-scale features or subdued features that are not obvious during field mapping may be discernable from afar. Soil sampling can also be of great importance, and analysis of the deposits may reveal the nature of the processes that led to their production, e.g. wind-blown sand is usually found to have a limited grain size distribution which can be correlated with the velocity of the prevailing wind. On the other hand, a marine fluviatile sediment may have a much wider grain size distribution and be characterized by rounded grains or a particular mineral assemblage.

Direct measurement of soil movement can be undertaken in many ways; for example, by placing soil strain probes which bend with the movement of the soil. For water- or wind-blown sediments, the sediment itself can be tagged by dyeing it, by introducing material of a different or distinctive colour, or by tracing the movement of naturally radioactive material or of material which has been irradiated artificially.

Wind erosion must also be considered. If reliable predictions cannot be made of future wind erosion potential, then engineered barriers should be considered as possible precautionary measures.

Geomorphology, in identifying surface features, can reveal past phases of faulting or uplift. Changes in drainage patterns or warping of erosion surfaces may also indicate upwarping or folding not necessarily connected with faulting.

Isostatic readjustment, which consists of the response of the earth's crust to the addition or removal of weight (e.g. ice), can also be correlated with changes in sea level or the positions of raised beach features. Conversely, subsidence features such as submerged forests can show the rate of crustal downwarping or of rise in sea level.

7.2.7. Hydrological investigations

The overall flow of water through an area requires a complementary study of surface-water hydrology and hydrogeology (groundwater hydrology). Both pathways are interrelated: water flows into a river from springs or seepages along its banks; or a river flowing over a permeable formation may drain into the ground and enter the groundwater regime.

7.2.7.1. Surface-water investigations

The characterization of the flow and changes in chemical and physical parameters of water at the surface of the earth is included in surface-water hydrology. Basic information required in surface hydrological studies includes precipitation, evapotranspiration and run-off, stream discharge, and measurement of changes in the storage of water in streams or lakes. The tracing of water currents within estuaries, lakes or the sea may also be involved in some cases.

Hydrological investigations should be interpreted in the light of regional and local topography, regional climate and local geology. These interpretations should be used in evaluating all areas under consideration at each step in the site selection process. When an area is being considered, relatively few stream discharge measurements on a few major streams can, with caution, be extrapolated to develop reasonable flood-frequency determinations for the general area. In many areas, long-term rainfall or stream flow, together with historical records of the incidence of flooding or drought, may be available.

Surface streams, lakes or seas are likely points of possible release of radio-activity from repositories into the human environment, and they can be the main agents for dispersion or dilution of such radioactivity. Surface hydrological processes will therefore be an important factor in the safety of a waste repository. Consequently, site-specific knowledge of the parameters discussed below will need to be determined:

(a) Precipitation

Precipitation includes rain, drizzle, snow and hail. There are numerous accepted instruments and techniques to measure its amount and intensity. Most are subject to errors caused principally by wind effects. Thus, it is important when relating rainfall figures for different localities to know the recording technique used and, if possible, to use similar instruments in similar surroundings or make appropriate corrections. The techniques that can be used depend on the range and type of precipitation to be expected and are well documented in the literature.

(b) Stream discharge

Stream flow data are required to define water input and output through an area. This information, together with changes in water storage, can be used to provide a water balance estimate.

Stream discharge can be determined by measuring the flow velocity at various positions over the cross-sectional area of the channel. Integration of these velocities over the area of the channel will give the total flow rate. Normally the height (or stage) of the water in a river or pond above a fixed point is measured and related to the discharge rate by a stage-discharge curve which can be determined empirically or from mathematical treatment of the shape of the channel.

The change in river stage with time (river hydrograph) is an important record, especially for determining groundwater contribution to stream flow at times of low rainfall. Numerous publications on methods of measuring stream discharge are available.

(c) Evapotranspiration

This parameter is not normally measured directly but is deduced by means of mathematical formulae developed from the knowledge of climatic parameters such as temperature, hours of sunshine, wind intensity, and the characteristics of soil and vegetative cover.

However, lysimeters (see Section 7.2.5) can be used to infer approximate evapotranspiration rates by the difference between rainfall, infiltration and soil moisture as follows:

Infiltration = rainfall − evapotranspiration ± changes in soil-water storage

Evaporation from open water surfaces is commonly measured using evaporation pans. These are small tanks of water in which measurements are made of the lowering of water level or the amount of water needed to maintain a constant level, taking into account replenishment by rainfall.

Evaporation from soil surfaces may be measured by percolation gauges. These consist of a cylindrical soil section enclosed in a buried metal container. A nearby rain gauge is used to measure rainfall on the cylinder and percolation from its base. Differences between input and percolation are taken as evaporation. To reduce the short-term effects of changes in storage, long-term records need to be known.

Measurements of transpiration from plants are extremely difficult to perform. Attempts have been made, using laboratory-based photometers and phytometers or, in the field, by placing a closed container or plastic tent over a plant and recording the resultant increase in humidity or the water vapour produced.

(d) Sedimentation

For long-term geomorphological studies it is necessary to determine the rate of erosion of a particular area. Since water erosion is so important in this context, measurement of the particulate load of a stream is used in order to relate it to erosion and deposition processes.

Many factors affect the rate of erosion, the most important of which are rainfall intensity, vegetation, soil type and land slope. All must be considered together. Sediment load in a stream is related to its stage, and can be subdivided into suspended sediment and bed-load, depending on where in the stream the material is being transported. Devices are available that recover sediment at different points in a channel. Numerous publications are available describing the techniques employed.

7.2.7.2. Groundwater investigations

Hydrogeological investigations are needed to characterize the potential movement of radionuclides in solution, away from the repository and into man's environment, usually at the earth's surface. This requires a detailed description of the extent of all geological formations (including clays and the more permeable aquifer materials, e.g. sandstone) and their physical and chemical properties. In addition, the hydraulic regime within the system must be understood.

Hydrogeological parameters relating to permeability, storage capacity and dispersion can only be obtained satisfactorily by field techniques invariably using boreholes. Much information can, however, be obtained from laboratory measurements on lithological samples carefully collected during drilling.

Drilling boreholes is appropriate to geological studies per se. Boreholes may also provide valuable information on the hydrogeology, and vice versa. Some of these techniques are discussed below.

7.2.7.3. *Field techniques for hydrogeological investigations*

Boreholes drilled to define some of the parameters listed in Table IV may also be used to perform tests to estimate various aquifer properties. They can be used for hydrogeological sampling and for observation points in pumping tests. Hydraulic properties of strata such as transmissivity and storage coefficients are basic to any hydrogeological study. Tests can be carried out by either injecting water into a hole or by pumping water out of a hole.

7.2.7.3.1. Pumping tests

The most efficient method for studying the fluid flow properties of aquifers is to carry out pumping tests in which water is pumped from a well at specified rates and the drawdown (i.e. rate of lowering) of the groundwater level in the pumping well and/or in one or several nearby wells is observed. Pumping tests can provide information on aquifer transmissivity, formation storage capacity, and leakage from semipervious layers above or below the aquifer being tested, and may also help identify the presence of recharge boundaries or boundaries which prevent groundwater flow. Pumping tests may last from a few hours to several weeks and require considerable organization if they are to be carried out successfully. To remove seasonal fluctuations in observed water levels, it is usually necessary to measure water levels for a week or so before and after a pumping test. Injection tests that generally require less time and equipment to perform may be carried out in single boreholes.

Most methods of analysing pumping test data require certain simplifying assumptions to be made about the transmissivity and storage properties of the aquifer so that mathematical formulae can be applied. Often the aquifer is considered to be isotropic (i.e. having the same physical properties in all directions), homogeneous (i. e. made of the same material with constant physical properties, on infinite areal extent), and groundwater flow is considered to be laminar.

In spite of these and other apparently restrictive assumptions, pumping tests, if carried out and interpreted with care, can provide sufficient information about an aquifer and its relation to adjacent geological formations to permit practical prediction of water flow rates.

7.2.7.3.2. Lefranc (or falling head) test

This is used to determine the permeability of those aquifers in which inter-granular flow occurs. It is accomplished by injecting water into a cavity of known dimensions at the bottom of a borehole and measuring the rate of disappearance of the water. The permeability of the rock can be calculated using standard equations.

7.2.7.3.3. Lugeon (or packer) test

This is used mainly to estimate the overall permeability of fractured rocks. It is carried out by injecting water at variable pressures between two packers (temporary seals) in the borehole and measuring the rate of disappearance of the water. The equivalent permeability of the rock is calculated using standard equations.

7.2.7.3.4. Slug test

This is a relatively rapid test to measure a limited portion of an aquifer. It consists in quickly withdrawing from or injecting a small discrete volume of water into a borehole and recording the rate of recovery or drop in water level in the hole, using a pressure transducer. Analytical methods to evaluate the permeability of the aquifer are then based on those from pumping tests.

7.2.7.3.5. Tracer tests

The artificial injection of tracers into a groundwater system can be used to determine groundwater flow rates and flow directions. This information can be used to determine the porosity and water permeability of a geological formation. Tracers can also be used to evaluate effects of dispersion under the prevailing hydraulic conditions of the period of observations.

Ideally, tracers should be easily detectable at low concentrations, should preferably not be present in the aquifer, and should not react with the soil or rock. They should also be non-toxic, and be miscible with and have a density near that of the groundwater. Tracers may range from simple salt solutions, easily detectable organic compounds or particulate material, to individual radioactive or non-radioactive isotopes. Organic tracers can usually be detected easily, but frequently they can react with the earthen media. Sodium chloride can be used in many cases because the chloride ion is usually not absorbed significantly.

Radioactive tracers are particularly useful, especially radioactive deuterium, tritium, iodine and chromium. Tritium produced from atmospheric testing of nuclear weapons since 1954 has been used to determine the residence time of water in an aquifer and to indicate the migration rate of water through the unsaturated zone by studying the vertical distribution of tritium with depth. Some of the different methods of using tracers are as follows:

The tracer is injected into the saturated zone, and groundwater is sampled in surrounding wells or springs for some time after injection.

The tracer is injected into a well and is then pumped back out of the same well (the 'alternate radial flow' method).

The tracer is injected into a central well, and samples are taken from surrounding wells (the 'radial divergent flow' method).

The tracer is injected into surrounding wells while pumping takes place continuously from a central well (the 'radial convergent flow' method).

In these tests it is important to choose carefully the location of the wells, and the distance between the injection well and the sampling point so as to ensure that a significant part of the aquifer is measured. Dispersion measurements, in particular, are related to the area covered by the investigation; small-area tests invariably give (erroneously) lower dispersion coefficients.

Tracers such as those naturally occurring in infiltrating water, e.g. ^{2}H, ^{3}H, ^{14}C or ^{18}O, can be used to study groundwater movement, origin and age. Tritium, with a half-life of 12.26 years, is particularly useful for indicating the fraction of recent groundwater infiltration in an aquifer. Tritium studies can also be of considerable use in estimating the rate of movement of moisture in the unsaturated zone (which otherwise is very difficult to measure directly). This technique requires the collection of samples at specific depths and the removal of the pore water in a high-speed refrigerated centrifuge. The tritium profile versus depth is then plotted to provide an index of the rate of moisture movement.

7.2.8. Hydrogeological transport of radionuclides

Before the suitability of a site for waste disposal can be assessed, the factors that can influence the potential migration of wastes in the hydrogeological environment must be understood and quantified. The final use for hydrogeological and geochemical information about the site is to evaluate the confining capacity of the geological barrier by modelling the radionuclide transport from the repository area to its potential outlets to the human environment.

It should, however, be remembered that although transport through the ground is generally considered to be the major pathway through which radionuclides may return to man, other pathways can be important, and these are considered in Section 8.

The principal factors to be considered in evaluating the hydrogeological characteristics of a site as a waste repository are:

(1) The leaching characteristics of the waste and the decay of radionuclides in the waste;

(2) The movement and interaction of waste with the rock or soil in the unsaturated zone; and

(3) The movement, dilution, dispersion and sorption of waste in the saturated zone.

Once a hydrogeological system has been characterized, the suitability of the natural features of a particular site and any additional (engineered) barriers that may be required to further retard waste migration can be assessed with respect to the potential movement of the radioactive constituents of the waste.

7.2.8.1. Leaching characteristics of the waste

Leaching characteristics of the waste form strongly affect the quantity of the radionuclides made available to leave the repository. The leached radionuclides are effectively the input to the predictive hydrogeological model. The leaching characteristics at a repository depend on many factors, including radioactive decay to new radioactive elements. The new element will then also be available to migrate according to its chemical properties. Studies based on expected waste constituents and their physical characteristics are necessary to determine:

The chemical interaction of radionuclides with other radionuclides or other non-radioactive wastes in the repository;
Corrosion or destruction rates of the packages in which the radioactive waste may be contained;
Desorption of radioactive substances from the waste form as a result of leaching by infiltrating water;
The form and physical characteristics of the material covering the repository;
The infiltration of water through the cover into the waste;
The method of waste emplacement and the expected level of operational quality assurance.

In general, the leaching rate is influenced by (a) temperature, (b) chemical composition of the leaching solution, and (c) the quality of the barriers between the waste and the water. Much of this information may be obtained by simulated field-leaching tests or by direct reference to similar installations. In practice, all ways of reducing infiltration of water into the waste or of reducing the mobilization of radionuclides with regard to possible waste interactions should be followed. For predictive purposes it is usually prudent to use conservative but realistic leaching rates. Leaching rates due to accidents or engineering failures will also need consideration. The characteristics of the leached materials could be expressed as follows:

The volume and composition of the leachate per unit time;
Radiochemical form and quantity of all important individual radionuclides in the leachate, per unit time.

These latter data may be obtained from radiochemical analysis and from spectrometric measurements of leachate samples under operational or simulated conditions and before contacting the groundwater.

7.2.8.2. *Migration of radionuclides in the unsaturated zone*

The equations governing flow of water in the unsaturated zone are complex but basically are considered to obey Darcy's law, where horizontal unsaturated flow, which is usually very limited, is neglected. One expression for the unsaturated flow rate in one dimension is:

$$V = -K\lambda \left[\frac{d\phi_z}{d_z} + 1 \right]$$

where

      V = flow through a given area in a given time
      K = saturated hydraulic conductivity
      λ is a term describing the change in hydraulic conductivity with moisture content

$\dfrac{d\phi_z}{d_z}$ = change in matric potential[6] with depth (i.e. the effective hydraulic gradient)

The λ term in this equation, which has a maximum value of 1 under saturated conditions, signifies that water movement in the unsaturated zone is vertical and occurs at a slower rate than under saturated conditions. Generally, the rate of water movement in fine-grained materials decreases rapidly by many orders of magnitude as the moisture content becomes less than that of saturation. The unsaturated zone can therefore play an important part in delaying the entry of radionuclides into the saturated zone of an aquifer.

During slow migration, nuclides and other waste constituents will come into intimate contact with the underlying or overlying geological formation, and a number of chemical reactions may take place leading to sorption. Common sorption mechanisms are ion exchange, where one atom in the solid is replaced by a different atom from the percolating solution; adsorption, where molecules are loosely bound onto the surface of minerals in the rock; precipitation and coprecipitation, where ions may come out of solution either by themselves or in conjunction with other minerals, usually because of a change in the physico-chemical environment. Physical factors such as filtration may effectively attenuate

---

[6] Matric potential is the suction pressure head in the unsaturated zone.

colloidal or particulate matter in a leachate, while biochemical mechanisms may take up various leachate components (especially nitrogen, potassium and phosphorus) and lead to a change in the hydraulic conductivity of the unsaturated zone due to biomass accumulation. Conditions that tend to mobilize the radionuclides, e.g. the presence of organic materials which may form soluble complexes, must also be considered.

Some factors that have been correlated with attenuation capacity in fine-grained materials are clay content (especially where the clay minerals have a high ion exchange or adsorption capacity), free content of hydrated oxides of iron and manganese and carbonate content (which stabilizes the pH). However, the prediction of migration through the unsaturated zone is difficult to achieve reliably and difficult to relate to measured parameters. Mathematical models are available but require further refinement. In general, it is often more precise to measure the migration rates directly by tracing the movement of a non-adsorbed chemical species or radioactive isotope such as ^2H, ^3H or ^{18}O through the unsaturated zone by drilling and analysing pore water from different depths at different times after introducing the tracer at the surface.

The natural hydraulic conditions under which flow has occurred are likely to change after a repository has been built. This change may result in increase or reduction in the rate of water movement, although a reduction should be expected in a well designed system.

7.2.8.3. Migration of radionuclides in the saturated zone

Radionuclide migration in the saturated zone is much easier to analyse and predict than unsaturated flow. The principal mechanisms affecting transport of a contaminant with the same density and viscosity as groundwater are:

(a) Convection

This is the movement of dissolved radionuclides with the average velocity of water. Its description requires the knowledge of the flow pattern within the aquifer, which can in turn be influenced by temperature.

(b) Dispersion

This is the combination of molecular diffusion that occurs under conditions of no flow and of solute movement due to the tortuosity of the different microscopic pathways in the ground that causes mixing and spreading in groundwater. This mechanism is characterized by a coefficient depending on the intensity of convection (or average pore-water velocity), and is referred to as the dispersion coefficient.

54

Dispersion coefficients depend on the area over which the investigation is made. If the dispersion coefficient is measured in a small part of the aquifer, effects such as increased dispersion caused by faults may not be evaluated correctly if the investigation is carried out in an unfaulted area.

(c) Interaction between nuclides and rock matrix

These interactions may be complex and may be caused by a number of physical and chemical or biochemical mechanisms which are interrelated and difficult to define individually. These basic phenomena are therefore not usually specifically distinguished in models; their total effects are usually empirically represented by a distribution coefficient, referred to as K_d, which assumes that there is a constant ratio between the amount of nuclides retained in rocks and the amount in solution. The general relationship can be written:

$$F = K_d C$$

where K_d is the distribution coefficient, C the concentration of radionuclides in water, and F the concentration of radionuclides in the solid phase.

This simple form of the distribution coefficient implies that the sorption phenomena are instantaneous and reversible, but research has shown that the rate of desorption of metals may be very much slower than sorption. However, these factors can be taken into account as the physicochemical kinetics are better understood. The use of the distribution coefficient also assumes equilibrium conditions which may not exist.

7.2.8.4. Mathematical modelling

Many methods have been proposed to solve the equation describing transport of radionuclides through a porous or fractured medium. All require specification of initial and boundary conditions for the flow system under study and of the release pattern of the radionuclide source. To illustrate one general form of the equation, consider a two-dimensional section through a flow system constructed parallel to the direction of flow, with no contaminant transport normal to the plane of the section. The general porous medium dispersion-convection equation, neglecting the effects of molecular diffusion but including a source-sink term for sorption and a radioactive decay term for dissolved constituents in a steady-state saturated flow system, can be written:

$$\frac{\partial}{\partial x_i} \left(D_{ij} \frac{\partial C}{\partial x_i} - V_L C \right) = R \frac{\partial C}{\partial t} + \lambda R C$$

where

    D_{ij} is the dispersion coefficient

    C is the concentration of the radionuclide in solution

    V_L is the average linear groundwater velocity through a unit section of the medium:

$$\left(V_L = -\frac{K_{ij}}{\theta} \frac{\partial h}{\partial x_j} \right)$$

    K_{ij} is the hydraulic conductivity

    θ is the effective porosity

    h is the hydraulic head

    R is the retardation factor $R = 1 + (\rho/\theta)K_F$

    ρ is the bulk density of the porous medium

    K_F is the distribution function, which is defined to be the distribution coefficient (K_d) if linear

    λ is the radioactive decay coefficient $(\lambda = 0.693/\text{half-life})$.

To solve this equation for a given flow system requires field data on the distribution of hydraulic conductivity, porosity, dispersion coefficients, hydraulic head, bulk density and the distribution function for various geological materials and geochemical conditions (including the effect of fractures), as well as the initial and boundary conditions and the contaminant release pattern. Analytical solutions for some one- and two-dimensional configurations with homogeneous material exist but, for most situations, numerical methods will be required. Several three-dimensional numerical flow models are now available.

7.2.8.5. Use of the results of hydrogeological modelling

To the extent limited by the accuracy of the model and the input data, the model can be used to compute the predicted distribution of the concentration of radionuclides migrating through the geological materials and the amounts of radioactivity reaching the outlets of the system. The computations can be used to simulate the repository in a variety of ways, including:

(a) To calibrate the model with experimental measurements. Part of this calibration may be done using piezometric data or the results of large-scale tracer tests performed on the site.

(b) To evaluate the sensitivity of the results to the different parameters. This will make it possible to determine the degree of confidence to be placed on the computations and will also show which parameters are important and need to be investigated more accurately.

56

(c) To simulate the behaviour of the repository under both normal and accidental conditions. These simulations can take different flow regimes into account.

The concentration of radionuclides reaching the points where they enter the human environment depends on the release rate from within the repository, and it is therefore important to characterize the role of the geological barriers by themselves. This can be estimated in the model by introducing the concept of attenuation into the model.

Let us suppose that the flow rate of any radionuclide at any given point of the repository is maintained at a constant value with time. Using this as a starting point, the model can estimate the increasing value of the radioactivity levels with time at the different repository outlets. The asymptotic value of these radioactivity levels with time obtained in this way at each repository outlet will represent the total attenuation effect inside the geological barrier for each respective outlet.

The attenuation factors can be computed for each radionuclide in the repository and can then be used as a source for transport through the human environment in the safety analysis, which is discussed in Section 8.

In general, the model is used to compute the estimated spacial distribution and concentration of radionuclides at varying times. In addition, it is used to estimate the maximum radionuclide level that may be released into the human environment under a range of conditions at the repository.

7.2.9. Geotechnical investigations

In constructing a shallow-ground repository it is unlikely that abnormal stresses will be applied to the ground unless extensive engineering works are to be undertaken to improve the natural geological barriers to waste migration. The main concern is likely to be the stability of the trench walls and that of the stockpile of material excavated from them. Geotechnical studies will therefore include the estimation of the safe angle of slope of the embankment so that there is no danger of collapse. Such studies are common in civil engineering practice and require a measurement of the cohesive properties of the formation either in the laboratory or in the field.

An important factor to be considered is the settlement of the waste in the trench after waste burial. If the final cover material(s) also subsides and punctures, then precipitation may allow rapid access of water to the waste and exacerbate the leaching process. This settlement will depend on several factors, including the type of waste, its rate of decomposition, and the compaction to which it is subjected before and after emplacement. Subsidence can only be measured satisfactorily by field experiments using similar wastes under similar emplacement conditions.

Other routine geotechnical tests include load-bearing, penetration, shear, and deformation tests in the field as well as a variety of laboratory tests on rocks as soil samples. If the excavated material is to be used as a final low-permeability cover, then the tests should be conducted to study the optimum compaction required.

7.2.10. Tectonics and seismicity studies

Evaluation of the tectonic and seismic risk (and possibly volcanic risk) is usually accomplished by analysing the tectonic characteristics of the area, taking into consideration all the available historical and instrumental earthquake data. The IAEA Safety Series No. 50, *Earthquakes and Associated Topics in relation to Nuclear Power Siting* [5], may be used for general guidance in estimating the size of a maximum expected earthquake over the next few decades. The estimated seismic response from such an earthquake could give some impression of the consequences to the strata and its stability at the repository site.

8. SAFETY ANALYSIS

Safety analyses and their assessments with criteria are important in every phase of system development: system selection; site investigation; repository design, construction, operation, shut-down and sealing; and the licensing processes relevant to these phases. Information from the site selection and confirmation activities is particularly vital to the successful safety assessment process and vice versa.

The safety analysis of a shallow-ground disposal facility requires an identification of the pathways by which radionuclides can enter and be transported to and through the human environment to man. A quantitative analysis then provides an evaluation of the potential individual radiation dose and collective dose commitment from normal and accident conditions at the facility. The results are then compared with the national regulatory limits and international recommendations for radiation doses. This section summarizes the principles and methods of calculating the radioactivity levels that may reach man, from which the dose to man can be calculated.

8.1. OBJECTIVES OF SAFETY ANALYSIS

The aim of a safety analysis for a proposed shallow-ground disposal facility is to calculate whether the potential radiation dose to man from normal and accident conditions is within the regulatory limits for acceptable doses. A safety analysis provides the major tool for assessing the technical suitability of a proposed waste disposal site and its engineered systems. Decisions at several stages in the site selection process will rely heavily on the results of safety analyses at a suitable level of detail. The final detailed site-specific analysis should demonstrate that the repository built on the site can comply with all safety criteria set by national authorities.

8.2. SAFETY ANALYSIS APPROACH

The overall relationships of the evaluations performed in safety analysis are discussed in IAEA Safety Series No.56, *Safety Assessment for the Underground Disposal of Radioactive Wastes* [8] and displayed in simplified fashion in Fig. 2. The evaluations usually start with basic definitions of the repository system and proceed through analyses of potential failure mechanisms. From here the transport of the radionuclides is assessed through the various compartments to result in a calculated source of radiation to man. These radioactivity sources are then used to calculate radiation exposures which are compared to regulatory values for acceptable radiation doses and to other criteria. This comparison and judgement on acceptability is termed *safety assessment*. If satisfactory, the study is complete; if not, modifications are made to the system (such as alternative sites, modified engineered feature, and/or more stringent waste characteristics) and the study is repeated. If probabilistic safety analysis is performed, additional evaluation of probabilities is required.

During the time that the buried wastes remain potentially hazardous, small amounts of radionuclides may be released under normal conditions from shallow-ground disposal facilities. The integrity of the disposal system may also be affected by unplanned events, which may generally be classified as natural and man-induced. A repository will usually be designed to withstand severe natural events anticipated during its life. Such a basis also provides a foundation on which to base the safety analyses; these should evaluate the consequences of normal or accident conditions for the entire period until the radioactivity of the total inventory in the repository decays to an acceptable level.

For an analysis of various radionuclide transport mechanisms to and through the environment, different mathematical models are used to express the relationships involved. The models for transport of radionuclides to and within the environment use as inputs information such as data on the waste source and

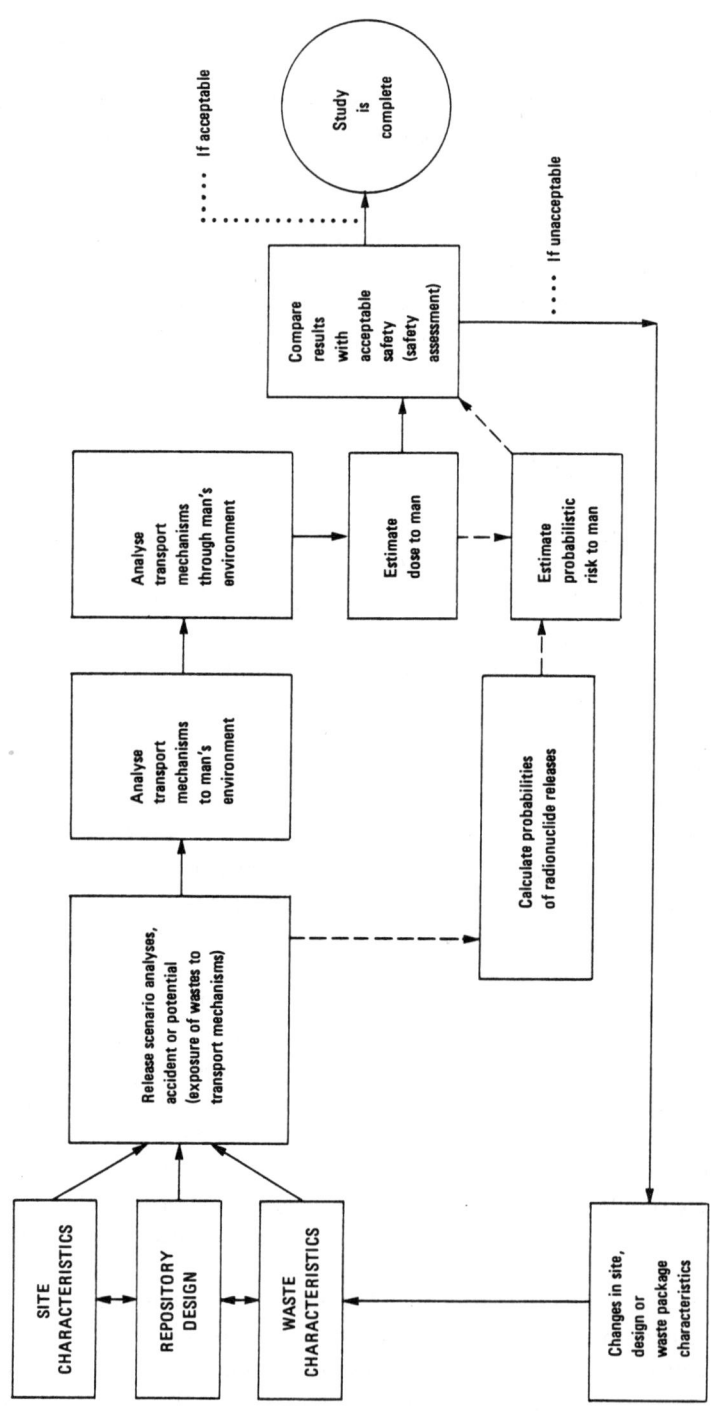

FIG.2. Relationships of safety analysis activities. Dashed line indicates additional activities when analysing probabilistic safety.

data on concentration, retention, dispersion, reconcentration and bioaccumulation of released radionuclides in and between each individual environmental compartment.

8.3. RELEASE SCENARIO ANALYSIS

Several types of events or processes could lead to the release of radioactivity to man's environment. To analyse radionuclide release scenarios, it is necessary to identify and define the relevant phenomena. Some of these phenomena could be due to the combined effects of waste and the repository (e.g. chemical, mechanical, radiological, etc.), the effects of endogenous natural forces (faulting, earthquakes, etc.), the effects of exogenous natural forces (erosion, meteorite impacts, etc.), and the effects of human or animal activity (drilling, burrowing, modification of geohydrology, etc.).

To identify phenomena that may be relevant to a particular radionuclide release scenario for a given repository design, it is useful to have a checklist, and such a checklist of the phenomena usually considered is presented in Table VI. It does not necessarily contain all phenomena relevant to shallow-ground disposal, and the divisions of the phenomena are somewhat arbitrary.

In principle, the following four approaches to radionuclide release scenario analysis may be used, either alone or in combination:

    (a)   Probabilistic safety analyses
    (b)   Deterministic analyses
    (c)   Stochastic analyses
    (d)   Qualitative analyses

These are discussed in Ref. [8].

8.4. CONSEQUENCE ANALYSIS

The consequence analysis begins by estimating the release of radionuclides from the waste form and ends with calculations of radiation dose to man. This evaluation should be carried out not only for the operational time of a facility and for normal conditions but also for the future (to the period when the radioactivity of the total inventory in the repository decays to an acceptable level) and for possible accident conditions.

Because of the special aspects of waste repository systems in shallow ground, and the expected performance of these repositories, it may not be feasible to demonstrate empirically their effectiveness for waste isolation over the required period. Analytical and mathematical modelling must therefore be used to predict the expected performance of the system. It is important that these

TABLE VI POTENTIAL CAUSES OF CHANGES IN SHALLOW-GROUND REPOSITORY SYSTEM

Natural processes or events:

Climatic change
Hydrology change
Sea level change
Denudation
Stream erosion
Glacial erosion
Excessive precipitation
Flooding
Sedimentation
Diagenesis
Diapirism
Faulting
Geochemical changes

Fluid interactions:
    Groundwater flow
    Dissolution
    Brine pockets

Uplift/subsidence:
    Orogenic
    Epeirogenic
    Isostasy

Magmatic activity:
    Intrusive
    Extrusive

Meteorite impact

Biological intrusion:
    Plants
    Animals

Man-made processes or events:

Poor site selection

Improper design and/or construction:
    Improper operation
    Improper waste emplacement

Operational accidents

Intentional intrusion:
    War
    Sabotage
    Waste recovery

Inadvertent future human alterations:
    Archeological exhumation
    Other waste disposal activities
    Resource mining (mineral, water, hydro-
      carbon, geothermal, salt, etc.)
    Irrigation
    Surface reservoirs
    Intentional artificial groundwater
      recharge or withdrawal
    Chemical liquid-waste disposal

Population distribution changes
Climate control
Large-scale alterations of hydrology

Waste and repository-caused processes or events:

Chemical effects:
    Geochemical alterations
    Waste package/geology interactions
    Corrosion of waste package
    Gas generation

Mechanical effects:
    Local subsidence
    Package movement

Radiological effects:
    Material property changes
    Radiolysis
    Decay-product gas generation

Thermal effects

Fluid movement changes

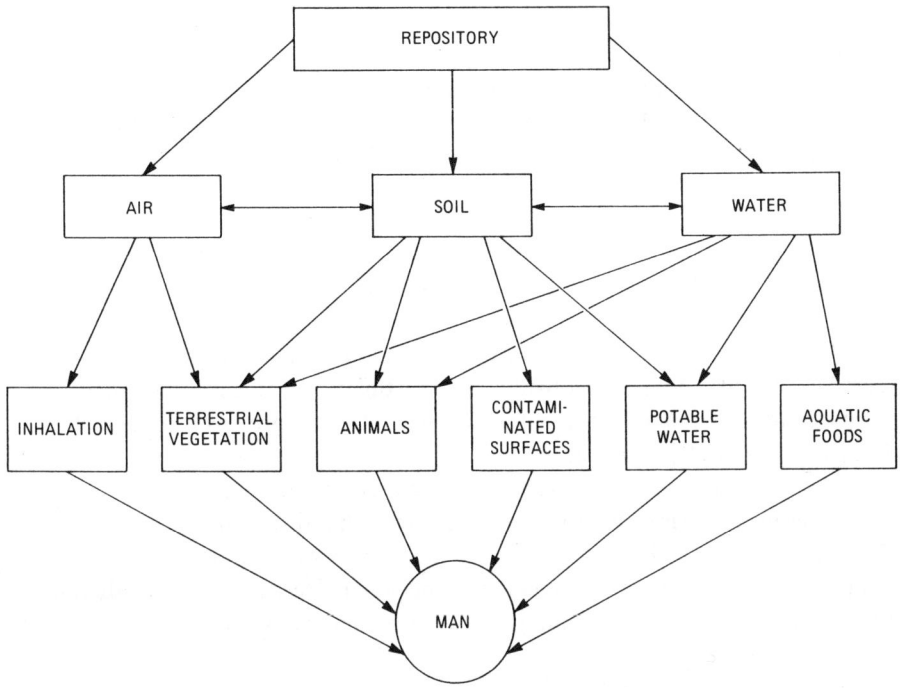

FIG.3. Simplified pathways of radionuclides to man from a shallow-ground repository.

models be validated by comparing their results with independent data obtained from field and laboratory studies.

In principle, there are two types of models that may be applied in different phases of the safety analysis: generic models based on general knowledge of the system and on data obtained from general literature reviews, and models specific to the site (these models and the data used therein are usually based on extensive field studies).

The complex relationships between the waste and the environment and within the environment offer a large number of pathways leading from the radioactivity source to man. The use of models normally implies considerable simplification of these pathways. This, however, may not impair their usefulness in the present context, provided the functions specifying the rates of transfer are properly chosen. Figure 3 shows a simplified network of the primary pathways of concern.

As indicated in Figs 2 and 3, potential pathways to be considered for radionuclides in the safety analysis are transfer from the repository into man's environment and transfer through man's environment to man. Such pathway analyses often involve analysis between several environmental compartments. Usually, only a few pathways are important in practice in a given geographical situation. This means that, for any given site, the large number of potential ways by which radioactive material can reach man can be reduced to a few pathways that are outstandingly more important than the others. These are referred to as the critical pathways.

8.5. EVALUATION AND APPLICATION OF RESULTS

After the dose consequences of potential radionuclide releases to man have been determined (and after probabilities are estimated, if this is to be done), the results should be evaluated by a safety assessment. This process is described in Ref. [8] and is briefly summarized.

The results of the safety analyses can be used by the appropriate bodies to make relevant comparisons and judgements, termed *assessments* in this report. The primary basis for assessment of the acceptability of a repository system is comparison with ICRP recommendations and other international and national criteria and regulations which apply. These assessments can be applicable to:

    (a) Repository concept evaluation
    (b) Repository site evaluation
    (c) Repository design evaluation
    (d) Repository system licensing

To supplement the comparison with criteria, or in the absence of specific criteria, comparisons can be made to put the results in perspective. Examples are:

    (a) Comparison of estimated probabilities and consequences of repository scenarios with the same quantities for other activities of the nuclear fuel cycle;
    (b) Comparison of estimated doses from repositories with doses for an equivalent period of time from naturally occurring radionuclides;
    (c) Comparison of calculated potential hazard against the hazard of naturally occurring ores;
    (d) Comparison of estimated effects of nuclear power generation with those of other power generation alternatives;
    (e) Comparison of the impact of nuclear wastes with hazardous chemicals in wastes from contemporary alternatives for power generation.

9. OTHER STUDIES

Some of the factors taken into account in site investigation for repositories in shallow ground are not related directly to safety as applied to earth sciences. These factors typically relate to the design, construction and utilization of the disposal facility and can be evaluated in economic or social terms.

9.1. REPOSITORY DESIGN

The design of a shallow-ground repository has a strong influence on site selection. The design for solid low- or intermediate-level radioactive wastes must take into consideration, in addition to factors usually involved in underground excavation and storage for other materials, the special requirements for confinement of waste for long periods. These may include the development of engineered barriers to limit radionuclide migration in order to supplement the barriers occurring naturally at the site. Engineered barriers can employ various physical and chemical principles to improve waste confinement, and it is therefore necessary to integrate the work of experts from many technical fields. Research and development work is frequently required for both natural and engineered barriers and the various processes that may affect them in the repository environment. This would often comprise theoretical studies and modelling, as well as laboratory work and field experiments. There will also be iterative interaction among site investigations, safety analyses and repository design.

In addition to considerations related to the long-term isolation of the waste, repository designs must incorporate the needs for safe operation of the repository while it is being constructed, while it is being filled with waste, and before it is backfilled and sealed. This generally involves application of well-known design principles for the containment of highly radioactive materials during handling. This is done to protect the repository workers and the public.

9.2. SOCIETAL FACTORS

Construction and operation of the repository may produce adverse impacts on the terrestrial environment and on the population in the area. The effects of noise, dust and human activity may, to a certain extent, disturb the ecological equilibrium of animal and plant species previously established. Siting of a repository may also cause adverse social response in the surrounding communities.

A short discussion of some of these factors follows. Other information may be obtained in IAEA Technical Report No. 177 [4].

9.2.1. Resource potential

Relevant information on mineral resources in the area should be collected and presented on maps of suitable scale. All areas that have been previously mined or are now undergoing active mining may present the potential for site-surface collapse, subsidence or uplift. This is also true of areas where similar man-made activities such as oil, gas or water extraction or injection have been performed. Such areas should be identified and evaluated for potential resources; if feasible engineering solutions do not exist for conflicts between the waste repository and the exploitation of the resource, the site may be deemed unsuitable and thus excluded from further consideration. For mineral resources not yet extracted, a study should be carried out to show that future mineral exploitation is not expected to conflict unacceptably with the use of the site as a repository.

9.2.2. Land value and use

The value of land is usually determined by its quality and by other factors such as distance from population centres, traffic conditions, etc. Evaluation of a repository site should involve careful assessment of the cost of the land, the transportation capabilities, the topography of the site and the location of major producers of radioactive wastes in the country.

An important factor may be the existing and potential use of land in the site area, e.g. for housing, agricultural, recreational and tourist purposes, or for ecological and archeological preserves. This factor may indeed have an important economic, societal and cultural significance. It is therefore recommended that studies be performed to ensure the integration of the repository and the potential use of land in its vicinity. These studies should ensure above all that:

(a)   All known competitive uses for land in the area have been considered;
(b)   Appropriate efforts have been devoted to minimizing all important potential negative impacts;
(c)   Plans for future development in the region are compatible with land uses around the site.

9.2.3. Population distribution

A potential repository site should be studied to evaluate the present and foreseeable future characteristics and distribution of the population of the region. Such a study should take into account any special characteristics that may influence the potential consequences of radioactive releases from the repository to the surrounding population. In cases where the repository could have a potentially significant effect on neighbouring countries or provinces, bilateral or international consultations should be undertaken.

9.2.4. Jurisdiction and rights of the land

To ensure control of the repository, including its surrounding buffer zone, it is highly desirable that the repository site, whether or not it is near a nuclear power plant/facility, be located on government-controlled land. Sites for nuclear power plants may have to be evaluated separately. The initial control of land by other authorities or private interests may introduce problems in land acquisition or human resettlement which should be considered as part of the overall site evaluation. Once a site is selected as a repository for radioactive waste, it should be ensured that its control is transferred to appropriate national government authorities. Consideration may also be given to the jurisdiction of the land after the repository can be released for unrestricted use.

Appropriate records should be reviewed to ascertain all existing rights, e.g. mineral rights, rights of way or easements on the proposed repository site. Each of these rights would have to be evaluated to determine whether exercise of the right would be incompatible with the safety of waste disposal. If possible interference were to exist, the impact of transferring the rights should be assessed.

9.2.5. Accessibility and services

Accessibility to a repository site for movements of personnel and materials during construction as well as the movement of operating personnel and radioactive waste during the waste emplacement phase is important. A site served by existing road and railway systems would be desirable. On the other hand, it may not be desirable to locate the repository near a heavily travelled transportation system because of the congestion that might result from the introduction of additional traffic. Alternatively, an improved transportation system constructed to serve the repository may improve transport conditions for local inhabitants and may encourage future development of industry or tourism in the region. Therefore, the capability of existing and potential transport routes should be studied with respect to the expected waste transportation requirements. The safety of transporting wastes to the area should also be evaluated.

The availability of electric power and water within reasonable commuting distance will provide, to some degree, a base for the necessary community services such as repository staff, housing and supporting services. The construction of additional services may create a favourable reaction on the part of the nearby population.

9.2.6. Other environmental impacts

Environmental impacts on or near the waste repository should be considered. The construction and operation of a proposed repository may result in a displacement of the ecological equilibrium for a given period. Although it is possible that the ecological equilibrium may be re-established after operational shut-down of the repository, a study to determine possible changes in animal and plant communities and the consequences should also be considered.

Other factors include the impacts on land, water and air use and quality, as well as noise and aesthetic impacts from the construction and operation of a repository. These impacts, however, are usually small for a shallow-ground waste respository.

9.2.7. Public attitudes

It is desirable to establish good relations with the population in the vicinity of the repository as well as with the public in general. Public support of an activity is highly desirable and is frequently required for the successful completion of a project. It is therefore advantageous to consider public information aspects at all population levels, national, regional and local, and at all stages of repository site investigations. In addition to exchange of information between experts and officials, it is important to include appropriate information exchange with members of the general public. Good communication concerning different activities among all affected groups should reduce conflicts and assure equitable distribution of benefits and liabilities.

To attain public acceptance or public support of the repository, it is desirable to locate, design, construct and operate the repository in such a manner that important adverse social impacts are kept to acceptably low levels, and that social benefits, such as generation of local employment, new housing and schools for the community, and improvement of local living conditions, are enhanced.

REFERENCES

[1] INTERNATIONAL ATOMIC ENERGY AGENCY, Underground Disposal of Radioactive Wastes: Basic Guidance, Safety Series No.54, IAEA, Vienna (1981).

[2] INTERNATIONAL ATOMIC ENERGY AGENCY, Treatment of Low- and Intermediate-Level Solid Radioactive Wastes, Technical Reports Series, IAEA, Vienna (in preparation).

[3] INTERNATIONAL COMMISSION ON RADIOLOGICAL PROTECTION, Recommendations of the International Commission on Radiological Protection, ICRP Publication 26, Ann. ICRP 1 3, Pergamon Press, Oxford (1977).

[4] INTERNATIONAL ATOMIC ENERGY AGENCY, Site Selection Factors for Repositories of Solid High-Level and Alpha-Bearing Wastes in Geological Formations, Technical Reports Series No.177, IAEA, Vienna (1977).

[5] INTERNATIONAL ATOMIC ENERGY AGENCY, Earthquakes and Associated Topics in Relation to Nuclear Power Plant Siting: A Safety Guide, Safety Series No.50-SG-S1, IAEA, Vienna (1979).

[6] ANNAN, A.P., DAVIS, J.L., "Radar sounding of bedrock and water table at Chalk River", Hydrological and Geochemical Studies in the Perch Lake Basin: A Second Report of Progress (BARRY, P.J., Ed.), Atomic Energy of Canada Ltd, Rep.No.AECL-6404 (1979) 123.

[7] SCOTT KEYS, W., MacGARY, L.M., Application of Borehole Geophysics to Water Resources Investigations (2nd edn), US Govt. Printing Office, Washington, DC (1972).

[8] INTERNATIONAL ATOMIC ENERGY AGENCY, Safety Assessment for the Underground Disposal of Radioactive Wastes, Safety Series No.56, IAEA, Vienna (1981).

BIBLIOGRAPHY

A. EARTH-SCIENCE INVESTIGATIVE TECHNIQUES

BARRETT, E.C., CURTIS, L.F., Introduction to Environmental Remote Sensing, John Wiley and Sons, New York (1976).

BEAR, J., Dynamics of Fluids in Porous Media, Elsevier, New York (1972).

DAVIS, S.N., DE WIEST, R.J.M., Hydrogeology, John Wiley and Sons, New York (1966).

DOBRIN, M.B., Introduction to Geophysical Prospecting (3rd edn), McGraw-Hill, New York (1976).

FREEZE, R.A., CHERRY, J.A., Groundwater, Prentice-Hall, Englewood Cliffs, NJ (1979).

FRIED, J.J., Groundwater Pollution: Theory, Methodology, Modelling and Practical Rules, Developments in Water Science, Vol.4, Elsevier, Amsterdam (1975).

STUMM, W., MORGAN, J.J., Aquatic Chemistry, Wiley-Interscience, New York (1970).

B. GENERAL

AGRICULTURAL INSTITUTE OF CANADA, Proc. Int. Conf. Land for Waste Management (TOMLINSON, J., Ed.), Ottawa (1973).

AVERY, T.E., Interpretation of Aerial Photographs, Burgess Publishing, Minneapolis (1977).

BOOTH, R.S., KAYE, S.V., ROHWER, P.S., "A systems analysis methodology for predicting dose to man from a radioactivity contaminated terrestrial environment", 3rd Natl. Symp. on Radioecology, Oak Ridge, 1971.

COMMITTEE ON GEOLOGIC ASPECTS OF RADIOACTIVE WASTE DISPOSAL, Division of Earth Science, National Academy of Sciences, Report to the USAEC (1966).

DUGUID, J.O., Status Report on Radioactivity Movement from Burial Grounds in Melton and Bethal Valleys, Oak Ridge Natl. Lab. Rep. ORNL-5017 (1975).

GARNER, R.J., Transfer of Radioactive Materials from the Terrestrial Environment to Animals and Man, CRC Press (1972).

INTERNATIONAL ATOMIC ENERGY AGENCY, Principles for Establishing Limits for the Release of Radioactive Materials into the Environment, Safety Series No. 45, IAEA, Vienna (1978).

INTERNATIONAL ATOMIC ENERGY AGENCY, Application of Safety Analyses for Repositories for Solid High-Level and Alpha-Bearing Radioactive Wastes in Continental Geological Formations, Technical Reports Series (in preparation).

INTERNATIONAL COMMISSION ON RADIOLOGICAL PROTECTION:

Recommendations of the International Commission on Radiological Protection, Committee II, ICRP Publication 2, Pergamon Press, Oxford (1959).

Recommendations of the International Commission on Radiological Protection, ICRP Publication 6, Pergamon Press, Oxford (1962).

Recommendations of the International Commission on Radiological Protection, ICRP Publication 10, Pergamon Press, Oxford (1968).

Recommendations of the International Commission on Radiological Protection, ICRP Publication 26, Pergamon Press, Oxford (1977).

INTERNATIONAL COUNCIL OF SCIENTIFIC UNIONS, SCIENTIFIC COMMITTEE ON PROBLEMS OF THE ENVIRONMENT, Environmental Impact Assessment: Principles and Procedures (MUNN, R.E., Ed.), SCOPE Rep. 5 (1975).

KENTUCKY DEPT. FOR HUMAN RESOURCES, Six Month Study of Radiation Concentrations and Transport Mechanisms at the Maxey Flats Area of Fleming County, Kentucky, State of Kentucky, Open-File Report (1974).

LeROY, L.W., et al., Subsurface Geology, Colorado School of Mines, Golden, Colorado (1977).

LOHMAN, S.W., Groundwater Hydraulics, Geological Survey, Washington, DC, USGS Prof. Paper 708 (1972).

LYON, R.B., RAMM – A System of Computer Programs for Radionuclide Pathways Analysis Calculations, Atomic Energy of Canada Ltd, Pinawa, Manitoba, Rep. AECL-5527 (1976).

NEW YORK DEPT. OF ENVIRONMENTAL CONSERVATION, Radioactivity in Air, Milk and Water, Jan.–Mar. 1975, State of New York, Environmental Radiation Bull. No.1 (1975).

NG, Y.C., BURTON, C.A., THOMPSON, S.E., TONDY, R.K., KAETNER, H.K., PRATT, N.N., Prediction of the Maximum Dosage to Man from the Fallout of Nuclear Devices, Pt.IV: Handbook, for Estimating the Maximum Internal Dose from Radionuclides Released to the Biosphere, Univ. of California Rep. UCRL-50163 (1968).

PAPADOPULOS, S.S., WINOGRAD, I.J., Storage of Low-Level Radioactive Wastes in the Ground – Hydrogeologic and Hydrogeochemical Factors, US Geological Survey, Open-File Report 74-344, prepared for US Environmental Protection Agency (1974).

PIRSON, S.J., Geologic Well Log Analysis, Gulf Publishing, Houston (1977).

POST, R.G., (Ed.), Proc. Symp. on the State of Waste Disposal Technology and the Social and Political Implications, 1979, College of Engineering, Univ. of Arizona, Tucson (1978).

RICHTER, D., LENNEMANN, W.L., "International aspects in geological disposal of nuclear waste", Storage in Excavated Rock Caverns (Proc. 1st Int. Symp. Stockholm, 1977) Vol.3, Pergamon Press, Oxford (1978) 723.

ROWAN, L.C., Application of satellites to geologic exploration, Am. Sci. 63 (1975) 393.

ROWE, W.D., HOLCOMB, W.F., The hidden commitment of nuclear waste, Nucl. Technol. 24 (1974).

SABINS, F.F., Jr., Remote Sensing: Principles and Interpretation, W.H. Freeman, San Francisco, (1978).

SCHLUMBERGER LTD, Log Interpretation: Principles, Vol.1, Schlumberger, New York (1978).

SCOTT KEYS, W., MacGARY, L.M., Application of Borehole Geophysics to Water Resources Investigations (2nd edn), US Govt. Printing Office, Washington, DC (1972).

SHERIFF, R.E., Encyclopedic Dictionary of Exploration Geophysics, Soc. Exploration Geophysicists, Tulsa (1973).

SIMMONDS, J.R., LINSLEY, G.S., JONES, J.A., A General Model for the Transfer of Radioactive Materials in Terrestrial Food Chains, Natl. Radiological Protection Board, Harwell, Publ. No. NRPB-R89 (1979).

SMITH, T.H., et al., A Risk-Based Fault Tree Analysis Method for Identification, Preliminary Evaluation, and Screening of Potential Accidental Release Sequences in Nuclear Fuel Cycle Operations, Battelle Pacific Northwest Labs Rep. BNWL-1959 (1976).

TELFORD, W.M., et al., Applied Geophysics, Cambridge Univ. Press (1976).

US DEPT. OF ENERGY, Interagency Review Group on Nuclear Waste Management, Report to the President, Techn. Inf. Doc. TID 29442, Washington, DC (1979).

US DEPT. OF ENERGY, Draft Environmental Impact Statement for the Management of Commercially Generated Radioactive Waste (2 vols), Rep. DOE/EIS-0046-0 (1979).

US DEPT. OF ENERGY, Statement of Position of the US Department of Energy in the Matter of Proposed Rulemaking on the Storage and Disposal of Nuclear Waste, Rep.DOE/NE-0007 (1980).

ZEMANEK, J., et al., Formation evaluation by inspection with the borehole televiewer, Geophysics **35** (1970) 254.

GLOSSARY

acceptable limit. Limit acceptable to the regulatory body.

aquifer. (i) A water-bearing formation below the surface of the Earth that can furnish an appreciable supply of water for a well or spring.
(ii) Geological formation or porous soil through which water may percolate for long distances, yielding groundwater to springs and wells.

barrier (natural or engineered). A feature which delays or prevents radionuclide migration from the waste and/or repository into its surroundings. An engineered barrier is a man-made or man-altered feature; it may be part of the waste package and/or part of the repository.

biosphere. That portion of the Earth's environment inhabited by any living organism. It comprises parts of the atmosphere, the hydrosphere (ocean, seas, inland waters and subterranean waters) and the lithosphere. The biosphere includes the human habitat or environment in the widest sense of these terms. (See **human environment.**)

buffer zone. A controlled area surrounding a nuclear installation (e.g. a waste repository) established to ensure an adequate distance between the installation and places used by or accessible to the public.

calcareous. Containing calcium carbonate.

clay. Minerals that are essentially hydrous aluminium silicates or occasionally hydrous magnesium silicates, with sodium, calcium, potassium and magnesium cations. Also denotes a natural material with plastic properties which is essentially a composition of fine to very fine clay particles. Clays differ greatly mineralogically and chemically and consequently in their physical properties. Especially because of their large surface areas, most of them have good sorption characteristics.

confinement (*or* isolation) of waste. The segregation of radionuclides from the human environment and the restriction of their release into that environment in unacceptable quantities or concentrations.

consequence analysis. A safety analysis that estimates potential individual and population radiation doses to humans, based on radionuclide releases and transport from a nuclear facility (e.g. a waste storage or disposal site) to the human environment as defined by hypothetical release and transport scenarios.

containment. The retention of radioactive material in such a way that it is effectively prevented from becoming dispersed into the environment or only released at an acceptable rate.

criteria. Principles or standards on which a decision or judgement can be based. They may be qualitative or quantitative.

decay, radioactive. A spontaneous nuclear transformation in which particles or gamma radiation are emitted, or X-radiation is emitted following orbital electron capture, or the nucleus undergoes spontaneous fission.

diapir. A piercement through geological strata in which a mobile core, such as rock salt, has injected into the more brittle overlying rock, generally forming geological folds or anticlines.

dispersion. The summed effect of those processes of transport, diffusion and mixing which tend to distribute materials from wastes or effluents through an increasing volume of water or air. The ultimate effect appears as a dilution of the materials.

disposal. The emplacement of waste materials in a repository, or at a given location, without the intention of retrieval.

dose equivalent. The product of absorbed dose and quality factor and all other modifying factors necessary to obtain an evaluation of the effects of irradiation received by exposed persons, so that the different characteristics of the exposure are taken into account. The special name for the SI unit of dose equivalent is the sievert (Sv); the rem may be used temporarily. (See ICRU Report 33.)

dose equivalent commitment (*or* effective dose equivalent commitment). For any specialized decision, practice or operation, the infinite time integral of the per-caput dose-equivalent rate for a specified population. The exposed population is not necessarily constant in numbers. It is commonly expressed in units of sieverts (Sv) or rems. (Note that this can apply over very long (geological) periods of time and that care must be taken to maintain perspective. The concept can be useful in making comparisons among alternatives but in a given case may have little meaning in an absolute sense.)

emplacement. Placing the waste in its location for storage or disposal.

endogenous. Originating from within. (See **exogenous**.)

engineered barrier, *see* **barrier**.

exogenous. Originating from without. (See **endogenous**.)

evapotranspiration. The sum total of water lost from the land by evaporation and plant transpiration.

fault. A fracture or zone of rupture in a rock along which there has been measurable displacement of the rocks on either side relative to one another, parallel to the fracture.

fission product. A nuclide produced either by fission or by the subsequent radioactive decay of the radioactive nuclide thus formed.

fold. A bend in strata or any geological structure.

formation. Any large and persistent assemblage of rocks which have some character in common, whether of origin, age or composition.

formation factor. The electrical resistivity of a rock saturated with an electrolyte divided by the resistivity of the electrolyte. The ratio of resistivities is a function of pore tortuosity, pore apertures and number of pores which relate to the porosity and permeability of the rock.

fuel cycle. All of the steps involved in supplying and using fuel materials for nuclear power reactors, including related waste management operations.

general (*or* generic) analysis. A generalized analysis for a nuclear facility; for a waste repository it is for a type of host rock, as opposed to an analysis for a site-specific host rock.

geochemistry. A science that deals with the chemical composition, the distribution of elements, and chemical changes in the crust of the Earth.

geohydrology. A science that deals with the properties, distribution and movement of water below the surface of the land (i.e. in the soil and underlying rock).

geological disposal, *see* **underground disposal.**

geology. A science that deals with the study of the Earth as a whole, its origin, its structure, its composition and history, and the processes which have given rise to its present state.

geomorphology. A science that deals with the description and interpretation of land forms.

geophysics. A science that deals with all the relevant physical phenomena that have a bearing on the structure, physical conditions and evolutionary natural history of the Earth. It includes the behaviour of earthquakes and the shock waves they produce, the gravity field and rotation of the Earth, the magnetic fields of the Earth, the temperature gradients and heat flows within the Earth, etc., and the capability of earthen materials to transmit or reflect these phenomena.

geotechnics. Dealing with the mechanical characteristics of rocks, e.g. stresses, strains and deformation of rocks under physical or thermal stresses.

groundwater. Water which permeates the (rock) strata of the Earth, filling their pores and cavities. (It excludes water of hydration.)

half-life, radioactive. For a single radioactive decay process, the time required for the activity to decrease to half its value by that process. (After a period equal to ten half-lives, the activity has decreased to about 0.1% of its original value.)

high-level waste. (i) The highly radioactive liquid, containing mainly fission products, as well as some actinides, which is separated during chemical reprocessing of irradiated fuel (aqueous waste from the first solvent extraction cycle and those waste streams combined with it).
(ii) Spent reactor fuel, it it is declared a waste.
(iii) Any other waste with a radioactivity level comparable to (i) or (ii). (Note that this definition is not related to "high-level waste unsuitable for dumping in the ocean" as used in the London Dumping Convention.)

host rock (*or* **host medium**). A geological formation in which a repository is located.

human environment *or* **man's environment.** Those portions of the Earth that are inhabited by man or readily available for use by man.

hydraulic conductivity. Ratio of flow velocity to driving force for viscous flow under saturated conditions of a specified liquid in a porous medium.

hydrogeology. The study of the geological factors related to the properties, distribution and movement of water below the surface of the land (i.e. in the soil and underlying rocks).

hydrology. The study of all waters in and upon the Earth. It includes underground water, surface water and rainfall, and embraces the concept of the hydrological cycle.

immobilization of waste. Conversion of a waste to a solid form that reduces the potential for migration or dispersion of radionuclides by natural processes during storage, transport and disposal.

intermediate-level (*or* **medium-level) waste.** Waste of a lower activity level and heat output than high-level waste, but which still requires shielding during handling and transport. The term is used generally to refer to all wastes not defined as either high-level or low-level. (See **alpha-bearing waste** and **long-lived waste** for other possible limitations.)

ion. An electrically charged atom or group of atoms.

ion exchange. A usually reversible exchange of one ion with another, either in a liquid or on a solid surface, or within a crystalline lattice.

isolation of waste, *see* **confinement of waste.**

isopach. A line on a map drawn through points of equal thickness of a geological formation.

joint. A fracture in rock, generally transverse to bedding, along which no appreciable rock movement has occurred on either side.

lithology. (i) The general characteristics of sediments, i.e. unconsolidated material forming sedimentary rocks.

(ii) The physical and mineralogical characteristics of rocks present in a stratigraphic subdivision, based on macroscopic features.

lithosphere. A broad, general term that refers to the upper rigid part of the Earth's crust. In a waste-management context it is used more loosely in describing storage and disposal practices which apply to the land as opposed to wastes discharged into the hydrosphere or atmosphere. The material composing upper parts of the lithosphere may be referred to as subsoil underlying a layer of soil as used in an agricultural sense. Occasionally the term 'soil' is found in reference to all forms of unconsolidated or semi-consolidated earth materials. An identifiable unit or stratum of material may be termed a rock.

long-lived waste. Waste that will not decay to an acceptable activity level in a period of time during which administrative controls can be expected to last. (See **short-lived waste.**)

long-term. In waste management, refers to periods of time which exceed the time during which administrative controls can be expected to last.

low-level waste. Waste which, because of its low radionuclide content, does not require shielding during normal handling and transport. (See **alpha-bearing waste** and **long-lived waste** for other possible limitations.)

migration. The movement of materials through a rock medium or some other solid substance, e.g. radionuclide migration.

model. In applied mathematics, an analytical or mathematical representation or quantification of a real system and the ways that phenomena occur within that system. Individual or subsystem models can be combined to give system models. Deterministic and probabilistic models are two types of mathematical models.

nuclear safety. A general term pertaining to the protection of people and property from the deleterious effects of radioactive contamination, exposure to ionizing radiation and a criticality excursion. (In this context, the term "ionizing radiation" may or may not include X-radiation produced by an X-ray machine, depending on national usage). (Also known as radiological safety.)

operation. All activities performed to achieve, in a safe manner, the purpose for which the facility was constructed, including maintenance, in-service inspection and other associated activities.

package, waste, *see* **waste package.**

permeability (of rock). The capacity of a porous or pervious rock for transmitting a fluid.

petrography. Pertaining to the systematic description and classification of rocks.

porosity. The ratio of the aggregate volume of interstices in a rock or soil to its total volume.

post-sealing period. The period after a waste repository has been shut down and sealed.

radioactive source term. The activities per unit time of radionuclide either leaving a nuclear installation or entering the environment or an environmental compartment.

radioactive waste. Any material that contains or is contaminated with radio-nuclides at concentrations or radioactivity levels greater than "exempt quantities" established by the competent authorities and for which no use is foreseen.

radioactivity. The property of certain nuclides of spontaneously emitting particles or of emitting gamma or X-radiation following orbital electron capture, or of undergoing spontaneous fission.

radiodecay heat. Heat generated by the absorption of radiation energy emitted by the decay of radionuclides.

radiolysis. Chemical decomposition by the action of ionizing radiation.

radionuclide migration. The movement of radionuclides through various media due to fluid flow and/or by diffusion.

recharge. Infiltration of water into the subsurface to become part of the ground-water in aquifers.

regulatory authority (*or* body). An authority or system of authorities designated by the Government of a Member State as having the legal authority for conducting the licensing process, for issuing licences and thereby for regulating the siting, design, construction, commissioning, operation, shut-down, decommissioning and subsequent control of nuclear facilities (e.g. waste repositories) or specific aspects thereof. This authority could be a body (existing or to be established) in the field of nuclear-related health and safety or mining safety or environmental protection, vested with such legal authority, or it could be the Government or a department of the Government, or it could be an international agency.

repository. An underground facility in which waste may be emplaced for disposal. The repository system includes the repository and all its supporting facilities.

rock. To the geologist any mass of mineral matter, whether consolidated or not, which forms part of the Earth's crust is a rock. Rocks may consist of only one mineral species, in which case they are called monomineralic, but they more usually consist of an aggregate of mineral species.

safety. Protection of persons and property from undue hazard (risk).

safety analysis. The analysis and calculation of the hazards (risks) associated with the implementation of a proposed activity.

safety assessment. A comparison of the results of safety analyses with acceptability criteria, their evaluation, and the resultant judgements made on the acceptability of the system assessed.

scenario analysis. Part of a safety analysis that identifies and quantitatively defines phenomena, their probabilities and their interactions, which could initiate and/or influence the release and transport of radionuclides from a source to humans. A release scenario defines the phenomena relevant to release of radionuclides from a radioactive (e.g. waste) source; a transport scenario defines the phenomena relevant to transport of the released radionuclides through the geosphere and biosphere to humans.

sedimentary rock. A layered formation of rock fragments laid down under water or land and usually subsequently cemented.

seismicity. Relating to vibrations of the Earth caused by earthquakes.

sensitivity analysis. An analysis of the variation of the solution of a problem with changes in the values of the variables involved. Two types of sensitivity analysis can be recognized. In simple parameter variation, the sensitivity of the solution is investigated for changes in one or more input parameters within a reasonable range about selected reference or mean values. In the perturbation analysis, the sensitivities of the solution with respect to changes in all input parameters can be obtained by applying differential and/or integral analysis.

shale. A laminated densely packed argillaceous sediment in which the constituent clay mineral particles are oriented parallel to the bedding planes.

shallow-ground disposal (*e.g.* **shallow-ground burial**). Disposal of radioactive waste, with or without engineered barriers, above or below the ground surface, where the final protective covering is of the order of a few metres thick. Some Member States consider this as a mode of storage rather than a mode of disposal.

short-lived nuclide. For waste management purposes, a radioactive isotope with a half-life shorter than about 30 years, e.g. ^{137}Cs, ^{90}Sr, ^{85}K and ^{3}H.

short-lived waste. Waste that will decay to a level which is considered to be insignificant from a radiological viewpoint, in a time period during which administrative controls can be expected to last. Such waste can be determined by radiological assessment of the storage or disposal system chosen. (See **long-lived waste**.)

shut-down and sealing. Action taken, after disposal operations have ceased, to prepare an installation for abandonment or minimal surveillance.

site. The area containing a nuclear installation (e.g. a waste repository) that is defined by a boundary and is under effective control of the implementing organization.

siting. The process of selecting a suitable site for an installation, including appropriate assessment and definition of the related design bases.

soil or rock mechanics, *see* **geotechnics**.

sorption. A broad term referring to reactions taking place within pores or on the surfaces of a solid. Its use avoids the problem of technical distinction between absorption and adsorption reactions.
Absorption is generally used to refer to reactions taking place largely within the pores of solids, in which case the capacity of the solid to absorb is proportional to its volume. *Adsorption* refers to reactions taking place on solid surfaces, so that the capacity of a solid is proportional to its effective surface area. An example of the latter is ion exchange, whereby ions occupying charged sites on the surface of the solid are displaced by ions from solution.

source term, *see* **radioactivity source term**.

stochastic event. A random event which can be predicted only by the probability of its occurrence. The term applies to data on phenomena that occur in time and/or space which are basically of a probabilistic nature but whose values depend partially on their respective time and/or space co-ordinates. In a stochastic time series, a term in the series is significantly related to the next one and this is considered in the time series analysis and synthesis.

stratigraphy. That branch of geology that treats stratified rocks and considers their formation, character, composition, deposition sequence in time, and correlation of different beds in the Earth's crust.

stratum (*or* bed). A layer of a geological formation consisting of approximately the same kind of rock material.

stress. Force applied per unit area of a solid.

subsidence. Sinking or caving of the ground surface.

surface water. Water which fails to penetrate into the subsoil and flows along the surface of the ground, eventually entering a surface drainage system.

tectonic. Pertaining to the rock structure, i.e. the external forms resulting from the deformation of the Earth's crust during the periods of mountain formation.

topography. (i) The configuration of (a portion of) the Earth's surface, including its relief and relative positions of its natural and man-made features.
(ii) The practice of graphical representation of the same.

transmissivity, hydraulic. Rate at which water is transmitted through a unit width of aquifer under a unit hydraulic gradient. It is expressed as the product of the hydraulic conductivity and the thickness of the saturated portion of the aquifer.

tsunami. Series of long, and in shallow water high, sea waves of great energy produced by submarine earth movement or by volcanic eruption.

underground disposal. Disposal of waste at an appropriate depth below the ground surface. This ground surface could be natural or artificially built up.

uplift. Rising up of the Earth's surface.

waste form. The physical and chemical form of the waste (liquid, in glass, in concrete, etc.) without its packaging.

waste management. All activities, administrative and operational, that are involved in the handling, treatment, conditioning, transport, storage and disposal of waste.

waste package. The waste form and any container(s) as prepared for handling, transport, storage and/or disposal. A container may be a permanent part of the waste package or it may be re-usable (shielding cask, shock absorbers, etc.) for any waste management step. The waste package may vary for the different steps in waste management.

water table. (i) The upper surface of the groundwater.
(ii) The upper surface of a zone of groundwater saturation.

well cuttings. Small fragments of rocks produced by the cutting or grinding action of a drill bit and returned to surface by the drilling fluid.

DRAFTING AND REVIEWING BODIES

LISTS OF PARTICIPANTS

1. Consultants' Meeting to draft the Working Paper, Prague, 23—27 July 1979

CZECHOSLOVAKIA

Dlouhy, Z.   Department of Radiation Protection,
                                    Nuclear Research Centre,
                                    250 68 Řež, Prague

FRANCE

Barbreau, A.   Département de sûreté nucléaire,
                                    Centre d'études nucléaires de Fontenay-aux-Roses,
                                    B.P. 6, F-92260 Fontenay-aux-Roses

UNITED STATES OF AMERICA

DeBuchananne, G.D.   Geological Survey,
                                    Department of the Interior,
                                    Reston, VA 22092

2. Advisory Group Meeting on Site Investigations for
   Repositories of Solid Radioactive Wastes in Shallow Ground,
   Vienna, 10—14 December 1979

AUSTRIA

Oszuszky, F.   Verbundgesellschaft,
                                    Am Hof 6, A-1010 Vienna

Gattinger, T.   Geologische Bundesanstalt,
                                    Rasumofskygasse 23, A-1030 Vienna

Krejsa, P.   Österreichische Studiengesellschaft für
                                        Atomenergie mbH,
                                    Forschungszentrum Seibersdorf,
                                    A-2444 Seibersdorf

Szeless, A.   Verbundgesellschaft,
                                    Am Hof 6, A-1010 Vienna

BELGIUM

Van de Voorde, N. *(Chairman)*

Centre d'étude de l'énergie nucléaire,
Boeretang 200, B-2400 Mol-Donk

Van Welden, K.

Conseiller au cabinet du ministre
de l'emploi et du travail,
51–53, rue Belliard, B-1040 Brussels

CANADA

Ophel, I.L.

Chalk River Nuclear Laboratories,
Atomic Energy of Canada Ltd,
Chalk River, Ottawa, Ontario K0J 1 JO

CZECHOSLOVAKIA

Dlouhy, Z.

Department of Radiation Protection,
Nuclear Research Institute,
250 68 Řež, Prague

EGYPT

Rassoul, A.

Atomic Energy Establishment,
101, Kasr El-Eini Street, Cairo

FRANCE

Barbreau, A.

Département de sûreté nucléaire,
Centre d'études nucléaires de Fontenay-aux-Roses,
B.P. 6, F-92260 Fontenay-aux-Roses

Berges, G.

Chargé de mission au secrétariat
général du comité interministériel
de la sécurité nucléaire,
27, rue Oudinot, F-75007 Paris

Cohen, P.

Office de gestion des déchets,
CEA, 29–33, rue de la Fédération, F-75752 Paris 15

Ledoux, E.

Ecole des mines de Paris,
35, rue Saint-Honoré, F-77307 Fontainebleau

INDIA

Balu, K.

Waste Management Operations Section,
Bhabha Atomic Research Centre,
Trombay, Bombay 400085

84

UNITED KINGDOM

Williams, G.M.

Environmental Protection Unit,
Institute of Geological Sciences,
Exhibition Road, London SW7 2DE

UNITED STATES OF AMERICA

Dressen, L.

Office of Nuclear Waste Management,
US Department of Energy,
Germantown, MD 20767

Tamura, T.

Oak Ridge National Laboratory,
Oak Ridge, TN 37830

DeBuchananne, G.D.

Geological Survey,
Department of the Interior,
Reston, VA 22092

INTERNATIONAL ATOMIC ENERGY AGENCY (IAEA)

Schneider, K.J. *(Scientific Secretary)*

Division of Nuclear Safety and
Environmental Protection,
IAEA, Vienna

3.   Revision of Draft Report, Vienna, 17–19 December 1979

Consultant

Williams, G.M.

Environmental Protection Unit,
Institute of Geological Sciences,
Exhibition Road, London SW7 2DE, England

4.   Technical Review Committee Meeting on the Underground Disposal of Radioactive Waste, IAEA, Vienna, 10–14 November 1980

Committee Members

BELGIUM

Heremans, R.

Géo-technologie du CEN/SCK,
Boeretang 200, B-2400 Mol-Donk

CANADA

Mayman, S.A.   Whiteshell Nuclear Research Establishment,
Pinawa, Manitoba ROE ILO

CZECHOSLOVAKIA

Malašek, E.   Czechoslovak Atomic Energy Commission,
Slezská 9, 120 29 Prague 2

FRANCE

Barbreau, A.   Institut de protection et de sûreté nucléaire,
CEN de Fontenay-aux-Roses,
B.P. No. 6, F-92260 Fontenay-aux-Roses

GERMAN DEMOCRATIC REPUBLIC

Runge, K.   National Board of Nuclear Safety and Radiation
Protection,
Waldowallee 117, DDR-115 Berlin-Karlshorst

GERMANY, FEDERAL REPUBLIC OF

Kühn, K.   Institut für Tieflagerung der Gesellschaft für Strahlen
und Umweltforschung mbH,
Berliner Strasse 2, D-3392 Clausthal-Zellerfeld

INDIA

Sunder Rajan, N.S.   High Level Waste Management Section,
Eng. Hall No. 5,
Bhabha Atomic Research Centre,
Trombay, Bombay 400 085

JAPAN

Doi, K.   Radioactive Waste Management Centre,
No. 15 Mori Building, 4F, 2-8-10 Toranomon, Tokyo

NETHERLANDS

Baas, J.L.   Ministry of Health and Environmental Protection,
P.O. Box 439, NL-2260 AK Leidschendam

SWEDEN

Larsson, A. *(Chairman)*   Swedish Nuclear Power Inspectorate,
                                  Box 27106, S-102 52 Stockholm

SWITZERLAND

Rometsch, R.   CEDRA,
                                  Parkstrasse 23, CH-5401 Baden

UNITED KINGDOM

Feates, F.S.   Department of the Environment, Becket House,
                                  1 Lambeth Palace Road, London SE1 7ER

UNITED STATES OF AMERICA

Vieth, D.   Division of Repository Development,
                                  US Department of Energy,
                                  Mail Stop B-107 (GTN), Washington DC 20545

Experts accompanying Committee Members

AUSTRALIA

Hardy, C.J.   Australian Atomic Energy Commission,
                                  Private Mail Bag, Sutherland, 2232, New South Wales

FRANCE

Barthoux, M.   Commissariat à l'énergie atomique,
                                  A.N.D.R.A., Agence nationale de gestion des déchets
                                      radioactifs,
                                  31–33 rue de la Fédération, F-75752 Paris 15

Berges, G.   Secretariat du comité interministerial de la sécurité
                                      nucléaire,
                                  27 rue Gudinot, F-75700 Paris

GERMAN DEMOCRATIC REPUBLIC

Noack, W.   National Board of Nuclear Safety and Radiation
                                      Protection,
                                  Waldowallee 117, DDR-115 Berlin-Karlshorst

JAPAN

Moriyama, N.   Reactor Safety Research Centre,
                                      Tokai Research Establishment, JAERI,
                                      Tokai-mura, Naka-gun, Ibaraki-ken

NETHERLANDS

Hamstra, J.   Netherlands Energy Research Foundation,
                                      P.O. Box 1, NL-1755 ZG Petten

SWEDEN

Boge, R.   National Institute of Radiation Protection,
                                      Box 60204, S-104 01 Stockholm

Rydell, N.   Swedish National Council for Radioactive Waste
                                          Management,
                                      Box 5864, S-102 48 Stockholm

UNITED KINGDOM

Grover, J.R.   AERE, Harwell,
                                      Oxfordshire OX11 0RA

International Organizations

COMMISSION OF THE EUROPEAN COMMUNITIES

Haijtink, B.   200 rue de la Loi,
                                      B-1049 Brussels, Belgium

NUCLEAR ENERGY AGENCY OF THE OECD

Gera, F.   38 boulevard Suchet,
                                      F-75016 Paris, France

INTERNATIONAL ATOMIC ENERGY AGENCY (IAEA)

Dlouhy, Z. *(Scientific Secretary)*   Division of Nuclear Fuel Cycle
Schneider, K.J. *(Scientific Secretary)* Division of Nuclear Fuel Cycle
Richter, D.K.   Division of Nuclear Fuel Cycle
Irish, E.R.   Division of Nuclear Fuel Cycle
Heinonen, J.U.   Division of Nuclear Fuel Cycle
Tsyplenkov, V.   Division of Nuclear Fuel Cycle

88

5. Final revision of the Report, Vienna, 17–21 November 1980

Consultant

Schneider, K.J.   Battelle-Pacific Northwest Laboratory,
                                    Battelle Boulevard,
                                    P.O. Box 999, Richland, WA 99352, USA

The following conversion table is provided for the convenience of readers

FACTORS FOR CONVERTING SOME OF THE MORE COMMON UNITS TO INTERNATIONAL SYSTEM OF UNITS (SI) EQUIVALENTS

NOTES:
(1) SI base units are the metre (m), kilogram (kg), second (s), ampere (A), kelvin (K), candela (cd) and mole (mol).
(2) ▶ indicates SI derived units and those accepted for use with SI;
  ▷ indicates additional units accepted for use with SI for a limited time.
  [For further information see the current edition of The International System of Units (SI), published in English by HMSO, London, and National Bureau of Standards, Washington, DC, and International Standards ISO-1000 and the several parts of ISO-31, published by ISO, Geneva.]
(3) The correct symbol for the unit in column 1 is given in column 2.
(4) ✳ indicates conversion factors given exactly; other factors are given rounded, mostly to 4 significant figures:
  ≡ indicates a definition of an SI derived unit: [ ] in columns 3+4 enclose factors given for the sake of completeness.

Column 1 Multiply data given in:	Column 2	Column 3 by:	Column 4 to obtain data in:	
Radiation units				
▶ becquerel	1 Bq	(has dimensions of s^{-1})		
disintegrations per second (= dis/s)	$1\ s^{-1}$	$\equiv 1.00 \times 10^0$	Bq	✳
▷ curie	1 Ci	$= 3.70 \times 10^{10}$	Bq	✳
▷ roentgen	1 R	$[= 2.58 \times 10^{-4}$	C/kg]	✳
▶ gray	1 Gy	$[\equiv 1.00 \times 10^0$	J/kg]	✳
▷ rad	1 rad	$= 1.00 \times 10^{-2}$	Gy	✳
▶ sievert *(radiation protection only)*	1 Sv	$[= 1.00 \times 10^0$	J/kg]	✳
rem *(radiation protection only)*	1 rem	$[= 1.00 \times 10^{-2}$	J/kg]	✳
Mass				
▶ unified atomic mass unit ($\frac{1}{12}$ of the mass of ^{12}C)	1 u	$[= 1.660\,57 \times 10^{-27}$	kg, approx.]	
▶ tonne (= metric ton)	1 t	$[= 1.00 \times 10^3$	kg]	✳
pound mass (avoirdupois)	1 lbm	$= 4.536 \times 10^{-1}$	kg	
ounce mass (avoirdupois)	1 ozm	$= 2.835 \times 10^1$	g	
ton (long) (= 2240 lbm)	1 ton	$= 1.016 \times 10^3$	kg	
ton (short) (= 2000 lbm)	1 short ton	$= 9.072 \times 10^2$	kg	
Length				
statute mile	1 mile	$= 1.609 \times 10^0$	km	
nautical mile (international)	1 n mile	$= 1.852 \times 10^0$	km	✳
yard	1 yd	$= 9.144 \times 10^{-1}$	m	✳
foot	1 ft	$= 3.048 \times 10^{-1}$	m	✳
inch	1 in	$= 2.54 \times 10^1$	mm	✳
mil (= 10^{-3} in)	1 mil	$= 2.54 \times 10^{-2}$	mm	✳
Area				
▷ hectare	1 ha	$[= 1.00 \times 10^4$	$m^2]$	✳
▷ barn *(effective cross-section, nuclear physics)*	1 b	$[= 1.00 \times 10^{-28}$	$m^2]$	✳
square mile, (statute mile)^2	1 mile^2	$= 2.590 \times 10^0$	km^2	
acre	1 acre	$= 4.047 \times 10^3$	m^2	
square yard	1 yd^2	$= 8.361 \times 10^{-1}$	m^2	
square foot	1 ft^2	$= 9.290 \times 10^{-2}$	m^2	
square inch	1 in^2	$= 6.452 \times 10^2$	mm^2	
Volume				
▶ litre	1 l *or* 1 ltr	$[= 1.00 \times 10^{-3}$	$m^3]$	✳
cubic yard	1 yd^3	$= 7.646 \times 10^{-1}$	m^3	
cubic foot	1 ft^3	$= 2.832 \times 10^{-2}$	m^3	
cubic inch	1 in^3	$= 1.639 \times 10^4$	mm^3	
gallon (imperial)	1 gal (UK)	$= 4.546 \times 10^{-3}$	m^3	
gallon (US liquid)	1 gal (US)	$= 3.785 \times 10^{-3}$	m^3	

This table has been prepared by E.R.A. Beck for use by the Division of Publications of the IAEA. While every effort has been made to ensure accuracy, the Agency cannot be held responsible for errors arising from the use of this table.

Column 1	Column 2	Column 3	Column 4
Multiply data given in:		*by:*	*to obtain data in:*

Velocity, acceleration

foot per second (= fps)	1 ft/s	= 3.048 X 10^{-1}	m/s $*$
foot per minute	1 ft/min	= 5.08 X 10^{-3}	m/s $*$
mile per hour (= mph)	1 mile/h	= $\begin{cases} 4.470 \times 10^{-1} \\ 1.609 \times 10^{0} \end{cases}$	m/s km/h
▷ knot (international)	1 knot	= 1.852 X 10^{0}	km/h $*$
free fall, standard, g		= 9.807 X 10^{0}	m/s^2
foot per second squared	1 ft/s^2	= 3.048 X 10^{-1}	m/s^2 $*$

Density, volumetric rate

pound mass per cubic inch	1 lbm/in^3	= 2.768 X 10^4	kg/m^3
pound mass per cubic foot	1 lbm/ft^3	= 1.602 X 10^1	kg/m^3
cubic feet per second	1 ft^3/s	= 2.832 X 10^{-2}	m^3/s
cubic feet per minute	1 ft^3/min	= 4.719 X 10^{-4}	m^3/s

Force

▶ newton	1 N	[≡ 1.00 X 10^0	m·kg·s^{-2}]$*$
dyne	1 dyn	= 1.00 X 10^{-5}	N $*$
kilogram force (= kilopond (kp))	1 kgf	= 9.807 X 10^0	N
poundal	1 pdl	= 1.383 X 10^{-1}	N
pound force (avoirdupois)	1 lbf	= 4.448 X 10^0	N
ounce force (avoirdupois)	1 ozf	= 2.780 X 10^{-1}	N

Pressure, stress

▶ pascal	1 Pa	[≡ 1.00 X 10^0	N/m^2] $*$
▷ atmosphere^a, standard	1 atm	= 1.013 25 X 10^5	Pa $*$
▷ bar	1 bar	= 1.00 X 10^5	Pa $*$
centimetres of mercury (0°C)	1 cmHg	= 1.333 X 10^3	Pa
dyne per square centimetre	1 dyn/cm^2	= 1.00 X 10^{-1}	Pa $*$
feet of water (4°C)	1 ftH$_2$O	= 2.989 X 10^3	Pa
inches of mercury (0°C)	1 inHg	= 3.386 X 10^3	Pa
inches of water (4°C)	1 inH$_2$O	= 2.491 X 10^2	Pa
kilogram force per square centimetre	1 kgf/cm^2	= 9.807 X 10^4	Pa
pound force per square foot	1 lbf/ft^2	= 4.788 X 10^1	Pa
pound force per square inch (= psi)^b	1 lbf/in^2	= 6.895 X 10^3	Pa
torr (0°C) (= mmHg)	1 torr	= 1.333 X 10^2	Pa

Energy, work, quantity of heat

▶ joule (≡ W·s)	1 J	[≡ 1.00 X 10^0	N·m] $*$
▶ electronvolt	1 eV	[= 1.602 19 X 10^{-19}	J, approx.]
British thermal unit (International Table)	1 Btu	= 1.055 X 10^3	J
calorie (thermochemical)	1 cal	= 4.184 X 10^0	J $*$
calorie (International Table)	1 cal$_{IT}$	= 4.187 X 10^0	J
erg	1 erg	= 1.00 X 10^{-7}	J $*$
foot-pound force	1 ft·lbf	= 1.356 X 10^0	J
kilowatt-hour	1 kW·h	= 3.60 X 10^6	J $*$
kiloton explosive yield (PNE) (≡ 10^{12} g-cal)	1 kt yield	≈ 4.2 X 10^{12}	J

---

^{a} atm (g) (= atü): atmospheres gauge
atm abs (= ata): atmospheres absolute

^{b} lbf/in^2 (g) (= psig): gauge pressure;
lbf/in^2 abs (= psia): absolute pressure.

Column 1 *Multiply data given in:*	Column 2	Column 3 *by:*	Column 4 *to obtain data in:*

Power, radiant flux

▶ watt	1 W	$[\equiv 1.00 \times 10^0$	J/s] ✻
British thermal unit (International Table) per second	1 Btu/s	$= 1.055 \times 10^3$	W
calorie (International Table) per second	1 cal$_{IT}$/s	$= 4.187 \times 10^0$	W
foot-pound force/second	1 ft·lbf/s	$= 1.356 \times 10^0$	W
horsepower (electric)	1 hp	$= 7.46 \times 10^2$	W ✻
horsepower (metric) (= ps)	1 ps	$= 7.355 \times 10^2$	W
horsepower (550 ft·lbf/s)	1 hp	$= 7.457 \times 10^2$	W

Temperature

▶ kelvin

$$\underline{K} \text{-- -- -- -- -- -- -- -- -- -- -- -- -- -- -- --}$$

▶ degrees Celsius, t $t = T - T_0$ ✻

     where T is the thermodynamic temperature in kelvin
     and T_0 is defined as 273.15 K

degree Fahrenheit	$\left.\begin{array}{l} t_{°F} - 32 \\ T_{°R} \\ \Delta T_{°R} \ (= \Delta t_{°F}) \end{array}\right\}$	$\times \left(\dfrac{5}{9}\right)$ gives	$\left\{\begin{array}{l} t \ \textit{(in degrees Celsius)} \ ✻ \\ T \ \textit{(in kelvin)} \quad\quad ✻ \\ \Delta T \ (= \Delta t) \quad\quad\quad ✻ \end{array}\right.$
degree Rankine			
temperature difference^c			

$$\text{-- -- -- -- -- -- -- -- -- -- -- -- -- -- -- --}$$

Thermal conductivity ^{c}

1 Btu·in/(ft^2·s·°F)	*(International Table Btu)*	$= 5.192 \times 10^2$	W·m^{-1}·K^{-1}
1 Btu/(ft·s·°F)	*(International Table Btu)*	$= 6.231 \times 10^3$	W·m^{-1}·K^{-1}
1 cal$_{IT}$/(cm·s·°C)		$= 4.187 \times 10^2$	W·m^{-1}·K^{-1}

Miscellaneous quantities

litre per mole per centimetre *(molar extinction coefficient or molar absorption coefficient)*	(1M/cm =) 1 ltr·mol^{-1}·cm^{-1}	$= 1.00 \times 10^{-1}$ m^2/mol ✻	
G-value, traditionally quoted per 100 eV of energy absorbed *(radiation yield of a chemical substance)*	1×10^{-2} eV^{-1}	$= 6.24 \times 10^{16}$ J^{-1}	
mass per unit area *(absorber thickness and mean mass range)*	1 g/cm^2	$[= 1.00 \times 10^1$ kg/m^2] ✻	

---

^{c} A temperature interval or a Celsius temperature difference can be expressed in degrees Celsius as well as
in kelvins.

HOW TO ORDER IAEA PUBLICATIONS

 An exclusive sales agent for IAEA publications, to whom all orders and inquiries should be addressed, has been appointed in the following country:

UNITED STATES OF AMERICA   UNIPUB, 345 Park Avenue South, New York, NY 10010

 In the following countries IAEA publications may be purchased from the sales agents or booksellers listed or through your major local booksellers. Payment can be made in local currency or with UNESCO coupons.

ARGENTINA	Comisión Nacional de Energía Atomica, Avenida del Libertador 8250, RA-1429 Buenos Aires
AUSTRALIA	Hunter Publications, 58 A Gipps Street, Collingwood, Victoria 3066
BELGIUM	Service Courrier UNESCO, 202, Avenue du Roi, B-1060 Brussels
CZECHOSLOVAKIA	S.N.T.L., Spálená 51, CS-113 02 Prague 1
	Alfa, Publishers, Hurbanovo námestie 6, CS-893 31 Bratislava
FRANCE	Office International de Documentation et Librairie, 48, rue Gay-Lussac, F-75240 Paris Cedex 05
HUNGARY	Kultura, Hungarian Foreign Trading Company P.O. Box 149, H-1389 Budapest 62
INDIA	Oxford Book and Stationery Co., 17, Park Street, Calcutta-700 016
	Oxford Book and Stationery Co., Scindia House, New Delhi-110 001
ISRAEL	Heiliger and Co., Ltd., Scientific and Medical Books, 3, Nathan Strauss Street, Jerusalem 94227
ITALY	Libreria Scientifica, Dott. Lucio de Biasio "aeiou", Via Meravigli 16, I-20123 Milan
JAPAN	Maruzen Company, Ltd., P.O. Box 5050, 100-31 Tokyo International
NETHERLANDS	Martinus Nijhoff B.V., Booksellers, Lange Voorhout 9-11, P.O. Box 269, NL-2501 The Hague
PAKISTAN	Mirza Book Agency, 65, Shahrah Quaid-e-Azam, P.O. Box 729, Lahore 3
POLAND	Ars Polona-Ruch, Centrala Handlu Zagranicznego, Krakowskie Przedmiescie 7, PL-00-068 Warsaw
ROMANIA	Ilexim, P.O. Box 136-137, Bucarest
SOUTH AFRICA	Van Schaik's Bookstore (Pty) Ltd., Libri Building, Church Street, P.O. Box 724, Pretoria 0001
SPAIN	Diaz de Santos, Lagasca 95, Madrid-6
	Diaz de Santos, Balmes 417, Barcelona-6
SWEDEN	AB C.E. Fritzes Kungl. Hovbokhandel, Fredsgatan 2, P.O. Box 16356, S-103 27 Stockholm
UNITED KINGDOM	Her Majesty's Stationery Office, Agency Section PDIB, P.O. Box 569, London SE1 9NH
U.S.S.R.	Mezhdunarodnaya Kniga, Smolenskaya-Sennaya 32-34, Moscow G-200
YUGOSLAVIA	Jugoslovenska Knjiga, Terazije 27, P.O. Box 36, YU-11001 Belgrade

---

 Orders from countries where sales agents have not yet been appointed and requests for information should be addressed directly to:

 Division of Publications
International Atomic Energy Agency
Wagramerstrasse 5, P.O. Box 100, A-1400 Vienna, Austria